VICTORY TO DEFEAT

OSPREY
PUBLISHING

RICHARD DANNATT
& ROBERT LYMAN

VICTORY
TO DEFEAT

THE BRITISH ARMY
1918–40

Dedication

This book is gratefully dedicated to the brave though benighted troops of the second British Expeditionary Force who were forced, by government parsimony and poor military preparation combined, to bear their country's burden in 1940.

It should never have been thus, if only their country had been awake, and had prepared adequately for the coming storm.

No treachery; but want of men and money,
Among the soldiers it is mutter'd,
That here you maintain several factions,
And whilst a field should be despatched and fought,
You are disputing of your generals;
One would have lingering wars with little cost:
Another would fly swift but wanteth wings:
A third man thinks, without expense at all,
By guileful fair words peace may be obtained.
Awake, awake, English nobility!

WILLIAM SHAKESPEARE
Henry VI Part 1 Act I, Scene 1

But if the watchman see the sword come, and blow
not the trumpet, and the people be not warned; if
the sword come, and take [any] person from among
them, he is taken away in his iniquity; but his
blood will I require at the watchman's hand.

EZEKIEL
33:6

It is not enough for those who love peace to talk
peace. A lover of peace must understand war – its
causes and its course. It is not enough to hope. We
must also work desperately on practical measures
that sometimes seem far short of our dreams.

JOHN G. WINANT
US ambassador to Britain, 1941–46

OSPREY PUBLISHING
Bloomsbury Publishing Plc
Kemp House, Chawley Park, Cumnor Hill, Oxford OX2 9PH, UK
29 Earlsfort Terrace, Dublin 2, Ireland
1385 Broadway, 5th Floor, New York, NY 10018, USA
E-mail: info@ospreypublishing.com
www.ospreypublishing.com

OSPREY is a trademark of Osprey Publishing Ltd

First published in Great Britain in 2023

ISBN: HB 9781472860866; PB 9781472860842; eBook 9781472860811;
ePDF 9781472860828; XML 9781472860859

23 24 25 26 27   10 9 8 7 6 5 4 3 2 1

Plate section image credits are given in full in the List of illustrations and maps (pp. 9–13).
Maps by www.bounford.com. The maps on pp. 33 and 138 were previously published in
1918: Winning the War, Losing the War (Osprey, 2018). The map on p. 287 was previously
published in *Case Red: The Collapse of France* (Osprey: 2017) and has been revised for this title.
Index by Alan Rutter

Typeset by Deanta Global Publishing Services, Chennai, India
Printed and bound in Great Britain by CPI (Group) UK Ltd, Croydon, CR0 4YY

Osprey Publishing supports the Woodland Trust, the UK's leading woodland conservation charity.

To find out more about our authors and books visit www.ospreypublishing.com. Here you
will find extracts, author interviews, details of forthcoming events and the option to sign up
for our newsletter.

Contents

List of illustrations and maps

with forward troops, carrying ammunition, fuel and rations to enable the advance to continue. (Photo by The Print Collector/Getty Images)

Their long war finally over, happy members of the Royal Field Artillery arrive in Dover following their return from Salonika, Greece in January 1919. (Photo by Historica Graphica Collection/Heritage Images/Getty Images)

Soldiers of the London Scottish Regiment march down Fleet Street towards Ludgate Circus during the Victory Parade. It was also hailed as the 'Peace Day Parade'. After such a long and terrible war it is unsurprising that few could stomach the thought that the world would ever contemplate another. (Photo by Central Press/Hulton Archive/Getty Images)

The Great War was an imperial affair. Here, Gurkhas of the Indian Amy Contingent march through Admiralty Arch to the Cenotaph during the Victory Parade. (Photo by Topical Press Agency/Hulton Archive/Getty Images)

Tanks, a potent symbol of British military power in Germany, being inspected at Düsseldorf by General Thomas Moreland, commander-in-chief of the British Army of the Rhine. (Photo by Daily Mirror/Mirrorpix via Getty Images)

British troops with an ad hoc form of vehicular mobility, equipped with a Lewis gun, on guard at the Jaffa Gate, Jerusalem, 1920. Their presence here was a legacy of General Allenby's successful campaign to oust the Turks during the campaign in 1917 and 1918. This occupation transited subsequently into the mandate for Palestine. (Photo by: Sepia Times/Universal Images Group via Getty Images)

From the Great War abroad to civil war at home. Two members of Sinn Féin are arrested following a raid on the Ministry of Labour offices in Dublin during the Anglo-Irish War (the 'Irish War of Independence') which broke out in its final incarnation in April 1916 and ended in July 1921. (Photo by Historica Graphica Collection/Heritage Images/Getty Images)

A detachment of British Lancers crossing the Rhine bridge into Cologne, Germany, part of the Allied occupation of the

Rhineland to enforce the Treaty of Versailles. (Photo by: HUM Images/Universal Images Group via Getty Images)

British troops leaving southern Ireland at the conclusion of the Anglo-Irish Treaty, signed in December 1921. (Photo by Independent News And Media/Getty Images)

In a spill over from the Great War, British troops found themselves in Gallipoli, guarding the road to Istanbul, during an attempt by Kemal Pasha to overturn the Great War settlement. The conflict, which began in 1919, was ended by the Treaty of Lausanne, signed on 24 July 1923. (Photo by Keystone-France/Gamma-Keystone via Getty Images)

Imperialism on the cheap. The RAF persuaded the government that recalcitrant tribesmen could be kept in their place by the use of aircraft dropping aerial bombs as a cost-saving measure. Here bombs await loading onto an RAF biplane somewhere in Iraq in 1923. (Photo by: Bristol Archives/Universal Images Group via Getty Images)

In an enduring military commitment to Egypt, an armoured car patrol makes its way along the Nile at the Old Cairo Quay. After its independence in 1922 Egypt recognised Britain's longstanding 'special interests' in the country, not least the Suez Canal, which Egypt promised to respect. (Photo by Hulton Archive/Getty Images)

Recruitment was a problem in the 1920s, only alleviated as a result of worsening economic conditions towards the end of the decade. This postcard from the Army Recruiting Stand at the British Empire Exhibition at Wembley in 1925 features a colour illustration of a contented pipe-smoking soldier. (Photo by Bob Thomas/Popperfoto via Getty Images)

In a classic Military Aid to the Civil Authorities operation, British troops deploy in London to protect convoys during the general strike of 1926. (Photo by: Photo12/Universal Images Group via Getty Images)

The British Army stationed 55,000 troops in India between the wars. Their task was to operate alongside the Indian Army in protecting the North West Frontier and offer Military Aid to the Civil Power. Here, troops rest after Hindu/Muslim communal violence in north Calcutta. (Photo by Topical Press Agency/Getty Images)

Men of the Scots Guards leave Southampton en route for service in China following the threat by Cantonese forces to British interests in the International Settlement at Shanghai. (Photo by Topical Press Agency/Getty Images)

A Royal Tank Corps Armoured Car Section patrols an area of the International Settlement in Shanghai. (Photo by Topical Press Agency/Getty Images)

A caricature of the British Army between the wars, neither one thing nor the other, uneasily managing the cusp between the old and new, in *Tatler*, No. 1369. The artist was Doris Zinkeisen (1897–1991). (DEA/ICAS94/Getty Images)

A posed photograph of men of the Royal Tank Corps on exercise on Salisbury Plain with a medium tank. The development of the tank, which few thinking soldiers disagreed with, was, however, delayed by confusion over the best way of deploying them in battle and by the lack of funds to develop them. (Photo by Fox Photos/Getty Images)

British troops return to the UK following the end of the post-war commitment to garrison the Rhineland. (Photo by Three Lions/Getty Images)

A British soldier in a Vickers Carden Loyd Machine Gun Carrier Mark VI tankette on a training exercise at Pirbright, Surrey in August 1929. (Photo by Fox Photos/Hulton Archive/ Getty Images).

Loyd carriers and horses on exercise together at Pirbright in August 1929, representing the nexus between old and new in warfare. (Photo by Fox Photos/Getty Images)

By December 1929 the British Army had withdrawn entirely from Germany. Here, men of the Royal Fusiliers march through Wiesbaden, the last of the British Army to leave the Rhine. (Photo by Central Press/Getty Images)

British troops at the Damascus Gate, Jerusalem, during riots in Mandated Palestine in August 1930. The commitment to

MAPS

Acknowledgements

The authors are indebted to all those who have contributed to bringing this book to fruition. It has taken several years of planning and was well in train before the Russian invasion of Ukraine gave our arguments – and efforts – a new emphasis, and urgency.

We wish first to thank a number of libraries and manuscript repositories in the United Kingdom, without which none of this would have been possible. Research and planning was made doubly difficult by Covid but we are grateful for their patience. We extend our thanks inter alia to the Keeper of the National Archives at Kew; the British Library; the Trustees of the Liddell Hart Centre for Military Archives at King's College, London; the Master, Fellows and Scholars of Churchill College, Cambridge University; the Controller of Her Majesty's Stationery Office and the Imperial War Museum for access to material and for permission to quote from the documents and manuscripts in their charge.

Many people assisted us in various ways as we tracked our way through the issues and arguments developed in these pages, some directly and others indirectly. Those who helped us indirectly did so by their writings, of which the full extent of our indebtedness can be found in the Notes and the Suggestions for further reading. For reason of the especial help provided by their own scholarship in the past we wish to acknowledge the work of the late Brian Bond; Brian Holden Reid; the late Corelli Barnett; the late Professor Trevor Wilson; Professor Nick Lloyd; Professor Lloyd

Clark; Professor Robin Prior; Professor Jonathan Boff; Professor Keith Jeffery; Adrian Phillips; Professor Gary Sheffield; Professor David French; Professor Ian Beckett; Professor John Charmley; Tim Bouverie; the late Professor Sir Michael Howard; Brigadier Allan Mallinson; Lieutenant General Sir John Kiszely; Dr Harold Winton; Dr Matthias Strohn and Walter Reid.

We are especially grateful to friends who have gone the extra mile by reading parts or all of the manuscript. To Brigadier John Powell; Professor Lloyd Clark; Professor Nick Lloyd and Lord Andrew Roberts we are indebted for helping us to develop our arguments, as well as from saving us from many errors and infelicities. It goes without saying that any remaining errors of judgement or fact are, of course, entirely our own.

We wish also to thank our agent, Charlie Viney, for steering this project to its conclusion, together with the advice, encouragement and skilful pen of our editor, Kate Moore; the senior desk editor, Gemma Gardner; the senior marketing executive, Elle Chilvers and the publicity executive, Rachel Nicholson, all of Osprey Publishing. And many thanks to Gilly Goldsmid who has sorted out many of Richard and Rob's diary issues.

Finally, we wish to acknowledge the steadfast support of our respective families during this endeavour. To our wives, Philippa and Hannah, both authors offer their grateful thanks, and love.

Introduction

La Forêt de Compiègne

On the early morning of 11 November 1918, when its usual occupants might have still been asleep, carriage 2419 – part of Marshal Foch's private train supplied by the Compagnie des Wagons-Lit – became the venue for the signing of the Armistice that ended the First World War. Between 5.12am and 5.20am, Marshal of France Ferdinand Foch, the Allied supreme commander, and Admiral Sir Rosslyn Wemyss, the British First Sea Lord, signed on behalf of the Allies. For Germany, four signatories were led by Matthias Erzberger, a civilian politician representing the new Social Democrat government in Berlin. The signing ceremony took place in a railway siding in a clearing in the Forest of Compiègne. The Armistice was to come into effect at 11am that morning as detailed to soldiers on the British front by an order from General Headquarters signed by Lieutenant Colonel William Dobbie, the duty staff officer. Two decades later he was to be the military defender of Malta. But at the time in 1918, so ended 52 tragic months of war when even on this final day of fighting there were 2,738 fatalities, the last of which was Henry Gunther, an American soldier killed some 60 seconds before 11am.

Fast forward to 22 June 1940. Into the same clearing swept a German delegation led by Adolf Hitler, in the company of Herman Goering, Wilhelm Keitel and Joachim von Ribbentrop. In the same railway carriage, which had become a monument to the

Allied victory of 1918, they demanded and received the surrender of France. The humiliation and revenge were completed when railway carriage 2419 was taken from the Forest of Compiègne to Berlin as an object of glee.

This book examines how the victorious Allies of 1918, including the largest army that Great Britain had ever put into the field, were defeated just 22 years later. How did this come about? And how did victory in 1918 for the British Army turn into defeat in 1940? Were the sacrifices made between 1914 and 1918 a pointless waste of lives?

Although it is popularly believed that the seeds of the Second World War were sown amongst the harsh terms of the Treaty of Versailles, this is an exaggeration, eagerly propagandised by the Nazi Party in its claims to victimhood. Admittedly Germany struggled desperately under the conditions imposed by the treaty, conditions which exacerbated the wretched nature of the German economy already impoverished by four years of Allied blockade. Although the war had been conducted on French and Belgian soil, the fragile nature of German society in 1918 was illustrated by a revolt by disenchanted groups of German sailors and soldiers in October. Yet just two decades later the next generation of German soldiers, sailors and airmen managed to defeat the Allies in a lightning campaign measured in days and weeks and not in months and years and placed Germany as the new masters of Europe.

This success was even more extraordinary given that in 1918 Germany had life-threatening problems in facing severe economic hardship. The war had cost it in excess of £45 billion and much of its population was in turmoil. Paying the price for victory had cost both France and Great Britain dearly too, not just the human cost in blood but the impact on national treasure as well – France spent over $24 billion during the war and Great Britain over $35 billion – creating unprecedented levels of indebtedness, not least to the United States of America. Indeed, the only country that seemed to benefit from the war in economic terms was the United States, but even this did not protect it from the Wall Street crash of 1929 and the Great Depression of the 1930s. But it is the response to these economic and social pressures that has the greatest relevance to explaining the reversal in fortunes in 1940. During the hardships of the 1920s and 1930s Germany's fragile democracy fractured, with Germany coming under the demagogic hand of the dictator Adolf Hitler. Although battered, the more mature democracies of France and Britain survived the social pressures that had swept away the tsar in Russia in 1917 and the Weimar Republic in Germany in 1933. Thereafter in the face of German territorial expansion, a new military build-up was always going to be faster in Germany, driven by dictatorial ambition, than in France and Britain hobbled by the lack of democratic consensus.

The main participants drew different strategic and military lessons from the First World War, especially from the closing campaigns of 1918. As the American Civil War had begun to show, military technology had developed to the point where high-performance weapons could inflict mass casualties on the contending armies. However, once the early weeks of manoeuvre in 1914 had not produced a decisive result for Germany, technology began to favour the defender, exemplified by the continuous line of fixed fortification trenches that ran from Switzerland through Europe to the North Sea. The impossibility of achieving a decisive breakthrough of these trench systems in the bloody battles of Ypres, Verdun, the Somme and Passchendaele led strategists to turn to geographical experiments, disastrously trying to knock Turkey out of the war through the ill-fated Gallipoli campaign. It was therefore left to innovative military and industrial minds to try to break the deadlock of the Western Front. The development of the tank, the aeroplane, more accurate artillery, better communications and much more intelligent tactical procedures ultimately proved decisive, but Germany and Great Britain took away different lessons from the latter months of 1918. The subsequent development of these lessons during the inter-war years produced two armies, very different in equipment, approaches and attitudes when they next faced each other in earnest in May 1940.

Another set of factors that seriously affected the performance of Germany and Great Britain in 1940 was their respective geopolitical circumstances. Germany, only a cohesive state since the days of Bismarck and his declaration of the German Empire in 1871, nevertheless was principally focused during the 1920s and 1930s on internal matters – first, the economy, then the rise of National Socialism and subsequently the ambition for territorial expansion predominantly in central Europe. Great Britain, on the other hand, had an extensive global empire stretching from Canada to Australasia to the Indian subcontinent and to the Far East. All these demands required money, manpower and leadership, at times deflecting political energy and attention away from the changing circumstances in Europe. The Indian independence movement was in its infancy but gathered strength with each passing year. The focus of the British Indian Army between 1919 and 1939 was on the border between India and Afghanistan, especially in the North West Frontier areas and Waziristan. Furthermore, as one of the Great Powers recognised by the League of Nations there were additional responsibilities placed on it. The Mandate for Palestine, shaped by the 1917 Balfour Declaration that promised a Jewish national home sitting in counterpoint to the motivation of the Arab Revolt and the possibility of an Arab state based on Transjordan, preoccupied London in a disproportionate way. Having defeated the Turks in Palestine in 1917 to 1918, it was inevitable that the British would have to take significant responsibility for filling the vacuum created by the collapse of the Ottoman Empire. Acceptance of the Mandate for Palestine in 1923 was an international burden that Britain would have to bear, albeit with some mildly duplicitous support from the French. Winston Churchill was to find himself in Palestine in 1921 and is credited with the near-disastrous establishment of modern Syria, Iraq and Jordan and the failure to satisfy either the Jews or the Arabs.

If overseas issues were not enough of a distraction, the immediate post-war British government of David Lloyd George also had to address the issue of Irish independence. A very live issue before 1914, the matter had gone largely onto the back burner during the First World War, with the marked exception of the Easter Rising of 1916. However, with the end of the war, the Irish Nationalists stepped up their campaign for independence, provoking some fairly extreme reactions from London. The Anglo-Irish Treaty of December 1921 gave independence to the majority of Ireland, which became the Irish Free State, while the six northern counties of Ulster remained under the Crown, requiring the renaming of the United Kingdom of Great Britain and Ireland as the United Kingdom of Great Britain and Northern Ireland in 1922.

As this book will relate, the British Army compounded the financial problems and overstretch of the decades that followed by taking its eye off the ball after 1918 with respect to the need to understand – and master – the nature of fighting a future highly intensive war. The ability to 'warfight' using all forces in a sophisticated, integrated and co-ordinated manner against a first-class enemy – one of the striking achievements of 1918 – was egregiously allowed to wither on the vine and ultimately to be forgotten by the British Army during the inter-war years.

Thus, the seeds for the defeat of the British Expeditionary Force (BEF) in France in 1940 were sown very liberally and widely in the two decades after the First World War. Once those seeds had taken root it became increasingly difficult for the tide of events to be turned. National Socialism strengthened and unified the German people. The view that the Great War had indeed been the war to end wars led to a grudging acceptance and appeasement of Germany's steady expansion in Europe into territories taken from the Central Powers at Versailles, unpicking the re-mapping of Europe. The remilitarisation of the Rhineland in 1936 was watched with dismay and, fatally, inaction, while the attempt to create 'Sudetenland' out of part of Czechoslovakia eventually provoked a belated response from Britain and France. This culminated in the Munich Crisis of September 1938, apparently settled by Neville Chamberlain obtaining Adolf Hitler's signature on a scrap of paper, waved proudly at the foot of the steps of his aircraft at Heston Aerodrome as he proclaimed 'peace in our time' on 30 September. It was not to be. Almost too late a period of rapid rearmament began in Britain, but it was a race in which Britain was dangerously behind Germany in respect of its thinking about and preparation for war. France, on the other hand, fearful of having to fight another war against Germany on its own, had decided in 1930 to place its faith in concrete and machine guns and accelerated the construction of the Maginot Line, even extending it further north towards the Channel ports from 1934.

The German and Russian invasion of Poland on 1 September 1939 began the final countdown to war. In March 1939, Britain had given assurances to Poland to protect its independence. London demanded that German operations cease by 11am on 3 September but the deadline passed with no such assurance from Germany. War was declared, Neville Chamberlain broadcasting to the nation at 11.15am. The second BEF that century was duly despatched to France and plans co-ordinated with the French for the defence of northern France and still-neutral Belgium. The grand strategy was to fight a long war, starving Germany of raw materials, mineral resources and food to the greatest degree possible in a re-run of the blockade of the Great War. No one was at all sure, however, whether Hitler really wanted to fight Britain and France again, especially not Britain for whom he seemed to have a grudging respect.

Events gathered pace. Russia attacked Finland, while France and Britain did nothing to help. Did Hitler misinterpret this as a signal that there was no fight in the stomach of the governments in either Paris or London? Step forward Winston Churchill, by then reappointed to his old post of First Lord of the Admiralty. If the strategy was to starve Germany of resources, then why not cut off supplies of iron ore from Sweden to Germany through Norway? His answer was first to mine the waters off Norway and then to secure the Norwegian ports through which the iron ore trade passed. This provided the strategic rationale for what turned out to be one of the worst operational decisions of the time: defeat, despite the heroism of the British soldiers involved, in the chaotic Norwegian campaign.

These were the circumstances that set the scene for the fateful debate in the House of Commons on 7 and 8 May 1940 that led to Churchill emerging as prime minister two days later, despite his involvement in yet another strategic disaster of Gallipolian proportions. The world had come full circle.

Prologue

Calais, 1940

Smoke drifted lazily above Calais as the overloaded *Kohistan* carrying the 2nd Battalion and all its vehicles and equipment staggered lumpenly through the waves on its final approach to the dockside. The smoke was the only sign that something was amiss. The morning was bright and clear. It was Thursday, 21 May 1940. Lieutenant Gris Davies-Scourfield of the 2nd Battalion, King's Royal Rifle Corps (the '60th Rifles'), awaited their arrival on the dockside with a mixture of anticipation and trepidation. This was where Britain had ordered the 700 men of his regiment, together with the Queen Victoria Rifles and the 3rd Royal Tank Regiment, to mount the empire's final defence. They were to do or die. The battle for Europe was nearing its humiliating finale, for France was surely lost and with it the fragile peace. Calais, lying eerily under the haze of drifting smoke, crowded with panicked, pushing refugees, was the end of the road for the British Expeditionary Force. The task of these units, constituting 30th Brigade, was to defend the port to allow an orderly evacuation of those remnants of the BEF able to make it to the coast. Even then its survivors were making their way in dusty columns across their fathers' battlefields to escape the enveloping panzers. It was clear to Davies-Scourfield when their vessel slipped into the port that a catastrophe of the greatest magnitude had overwhelmed the French and British armies. Only the last-ditch defence of the Channel ports would hope to prevent the army's annihilation.

Three hundred thousand men were in full retreat, fighting their way back to salvation at the hand, they hoped, of the Royal Navy. The 60th Rifles were told that 'weak enemy detachments' were making their way towards Calais from the south. These forces had already engaged the Territorials of the Queen Victoria Rifles and the Cruiser tanks of the 3rd Royal Tank Regiment – which had arrived in Calais the previous day – that morning. Pulling slowly into Calais harbour to unload, the men were shocked by the sight of military refugees flooding into the town. A lone German aircraft weaved across the sky, dropping a bomb on the port, its jagged path followed by little black bursts of anti-aircraft fire. A hospital train disgorged its pathetic cargo of dead and wounded. Davies-Scourfield had never seen a dead body before. An elderly-looking major bumped into him on the dockside: 'Come on, young feller, you'd better get away while the going is good; there's only one safe place just now, and that's England. This may be your last chance.'[1]

Without dock labour it took three days of unrelenting effort to extract the battalion's vehicles and equipment from the hold of the ship, all the while under German artillery bombardment. Meanwhile, it was clear that a sacrifice was intended. Antony Eden, Secretary of State for War, sent the troops a signal that made the requirements for sacrifice unequivocal:

> Defence of Calais to utmost is of vital importance to our Country and BEF and as showing our continued cooperation with France. The eyes of the whole Empire are upon the defence of Calais and we are confident you and your gallant regiments will perform an exploit worthy of any in the annals of British history.[2]

Davies-Scourfield's task was to defend a piece of high ground to the south of the port on the road to Boulogne. The 'weak enemy' which faced them was in fact the whole of Major General Ferdinand Schaal's 10th Panzer Division, supported by an additional 1st Panzer Division and all of the artillery of General Heinz Guderian's XIX

Panzer Corps. Digging in was easy, he recalled, and his defensive position would have been formidable if his men had had some wire, mines and a weapon able to defeat a German medium tank. Moreover, large gaps existed between their defences. There simply weren't enough men to go around.

Sudden surges of refugees coming into their position, military and civilian, gave notice of the advancing enemy tanks. A patrol of riflemen and three Bren gun carriers that he sent out ahead to scout for enemy activity were chewed up by tank fire. Two of the carriers and their crews were destroyed. The commander of the third managed to get back with the news, the Boys anti-tank gunner dead in his seat. A constant smattering of incoming artillery that morning showed something of Schaal's intentions. An assault across open ground to his front by German infantry was turned back by artillery fire, but the German advance was relentless. By noon on 24 May the German infantry began to advance, accompanied by tanks, artillery fire and machine guns. Stuka dive-bombers dropped from the sky with the terrifying wail of their sirens preceding the inevitable arrival of their bombs. All the while artillery fell among their positions. Some tanks acted as mobile pillboxes, shooting up the British positions, while others, together with infantry, worked their way around the flanks. The men of the 60th Rifles, conscious that they did not have limitless ammunition, replied carefully, attempting to make the Germans unwilling to enter into close combat. Two 2-pounder anti-tank guns were quickly destroyed by a German tank. The infantry's Boys anti-tank rifles hit their targets many times, but never did much more than ricochet off the German armour. They had been obsolete long before they were even issued to the troops, one in a line of dismal pre-war procurement decisions. Davies-Scourfield was thankful for the battalion's 3-inch mortars, however, which produced a good killing of unprotected German infantry. Nine antiquated Swordfish biplanes of the Royal Navy helped by dropping bombs on the advancing enemy tanks on 27 May. They were lucky not to have encountered any of the Messerschmitts which routinely accompanied the German Stukas into battle to protect them from a loitering Spitfire or Hurricane.

He saw a high-winged Lysander spiralling to the ground with one of its wings shot off.

At about 8pm on the first day of battle the German attacks were called off. Davies-Scourfield and his men, reduced from 40 that morning to about 25, picked themselves up and retired into Calais proper. They had neither the men, nor defensive equipment nor ammunition to fight in their forward positions for another day. Their task was now to hold Calais from within the town itself, occupying houses flanking the canal. The truth was – and they knew it – that there was no way out. Evacuation was out of the question. The defence of Calais was to be the start and the end of their war. Theirs was to be the last-ditch stand of the BEF, the forlorn hope of a fighting campaign that had lasted only a matter of weeks, most of it entailing an exhausting withdrawal from Belgium to the Channel ports.

PART ONE

The Great War

Chapter 1

The *Kaiserschlacht* and its consequences

On 21 March 1918, *Kaiserschlacht*, the 'emperor's battle', began. It was to be the last throw of the dice, designed to bring the war to a close in Germany's favour. Operation *Michael*, the last major German offensive of the Great War, began at 4.40am with a five-hour artillery bombardment so powerful it could allegedly be heard in London. At 9.40am, infantry began their advance behind a massive creeping barrage across a 46-mile front. Led by groups of specially trained assault troops known as 'storm troopers', some 67 German divisions in three armies, many of which had recently arrived from the Eastern Front following the withdrawal of Russia from the war, began their advance. Their operational design was to split the British armies from the French and to penetrate deep into the British rear areas forcing them to contemplate evacuation from France. Thereafter they would focus on destroying the French armies, thought likely to collapse without the support of the British, and thus win the war before the recently arrived United States Army could make a decisive intervention.

Initially the German attack went well, assisted by the thick fog that hung over the battlefield. The British Fifth Army under General Hubert Gough – on the British right or southern flank – had a well-manned first line of defence based on a series of strong points. These suffered terribly from the opening bombardment. The new defensive arrangement based on mutually supporting strong points rather than in a continuous line made perfect sense but meant that German assault groups could bypass many of the forward British positions leaving them to be dealt with by follow-up forces, with the storm troopers striking towards the rear areas and thereby causing command consternation and disarray. By the end of the first day, although the Germans had only penetrated the British second line of defence in two places, the British chain of command was becoming concerned. By nightfall on 21 March, Field Marshal Sir Douglas Haig was already worried about the security of the Channel ports even though he had deployed most of his reserve divisions to his northern sector to protect them. Haig had assumed that the French would come to General Gough's aid in the south, as had been agreed. However, the French Army commander on Gough's right, General Philippe Pétain, believed that the attack on Gough was only a feint. He thought that the main German thrust would come even further south on the axis Rheims–Paris. As a result, he was reluctant to move his reserves north to help Gough. Thus, the German Eighteenth Army by the end of 21 March appeared to have had the most success even though it was only making a supporting attack in the overall German design for battle.

The Germans then had an operational dilemma – to reinforce Eighteenth Army's relative tactical success on the first day or to persist with the main effort attacks by Second Army, whose mission was to strike west towards the Channel ports. Contrary to received German doctrinal wisdom, General Erich Ludendorff, in overall command of Operation *Michael*, made the fateful decision to switch the point of main effort from Second Army to Eighteenth Army. Not only was this very difficult to do, as the weight of German logistical support had been stockpiled to support Second Army, but this critical decision effectively meant that the operational design was abandoned and with it the chance of strategic success. The significance of this German decision was not appreciated by the British, embroiled as they were in the fog of war. As a result, three days later plans began to be developed to pull Gough's Fifth Army back towards the Channel ports and, as a further measure of the pressure felt, Haig agreed at a meeting of the Supreme War

Council on 26 March to place the BEF under the command of General Ferdinand Foch for the duration of the battle. This turned out to be a sound and sensible decision that should have been taken much earlier in the war. The German offensive continued until 5 April when Ludendorff ordered the suspension of further attacks.

Despite seizing ten times as much territory in 16 days as the British had seized in the three months of the Battle of the Somme in 1916, the German offensive had failed with staggering losses. Casualties at around a quarter of a million each had been much the same for both Germany and the Allies, but the German Army had largely exhausted itself and reserves were no longer plentiful. Moreover, it was now dangerously exposed in a bulge centred on St Quentin. Critically, the offensive had been halted nine miles short of Amiens, a city of major significance to the British and through which most of its supplies passed between the Channel ports and the front. The loss of Amiens would have been a major communication and logistic headache for Haig. By 5 April, in contrast to the Germans, the British Army, while battered, was not broken. Planning for counter-offensive moves began immediately.

The significance of the German Spring Offensive, which continued until the final attack on 15 July, is less that the overall offensive was the last chance Germany had to win the war, but more in the lessons that the opposing armies drew for future conflict. For both sides, artillery had become more accurate and sophisticated, significantly helped by improved battlefield communications. Short, sharp preparatory bombardments followed by creeping barrages replaced the week-long attempts to destroy everything in front of an attacking force that had characterised the British battles of 1916 and 1917. Moreover, control of the airspace above the battlefield was seen to be a major force multiplier both in terms of reconnaissance and of controlling artillery fire. However, the opposing sides learned different lessons about the integration of command and control of air and land forces. In the midst of the battle the long-planned formation of the Royal Air Force took place on 1 April, the British Army losing its integral Royal Flying Corps. Above the battlefield the change itself did not affect tactical activity, although over time the formation of a separate service would introduce tensions over doctrine and resources that had hitherto not existed. However, on the German side there remained a clear understanding of the need for close integration between air and land forces, a feature that would be so dramatically demonstrated in May 1940.

The Germans also realised that specially trained and equipped fighting groups could restore manoeuvre and momentum to the battlefield. The ability of storm trooper groups to outflank, infiltrate through and bypass strong points illustrated the principle of probing everywhere and then reinforcing success at identifiable weak points. At the tactical level this worked brilliantly until the groups of specially trained storm troopers were either exhausted or killed. The lesson the Germans took from this was that for manoeuvre to succeed it must be rapid and sustainable and by vehicle, not just on foot. Their major mistake in March and April 1918 was to allow the exploitation of tactical success to have an undue bearing on the achievement of the operational plan. Ludendorff's design for battle had been as good as it could have been in the fourth year of an exhausting war, but he needed to reinforce his primary attack – Second Army in the centre – rather than switch his main effort. In the next war, General Heinz Guderian did not make the same mistake when he famously argued: 'Klotzen, nicht Kleckern!' – 'Clout, don't dribble!' If Second Army had been reinforced to 'clout' properly, the course of the Great War could have been very different.

The final German attack on the Western Front took place on 15 July aiming to capture the city of Rheims and open up the route to Paris, as General Pétain had anticipated. Although Ludendorff's original operational design had been compromised back in April, he still managed to assemble 42 divisions and nearly 1,700 guns for this further attack. It failed, however, within its first few hours. The Allies had amended their defensive tactics, holding the forward lines very lightly, ready to strike back with significant counter-attack forces. It was then that the German nemesis revealed itself – the United States Army with four large divisions, supported by

Operation *Michael*, March–April 1918

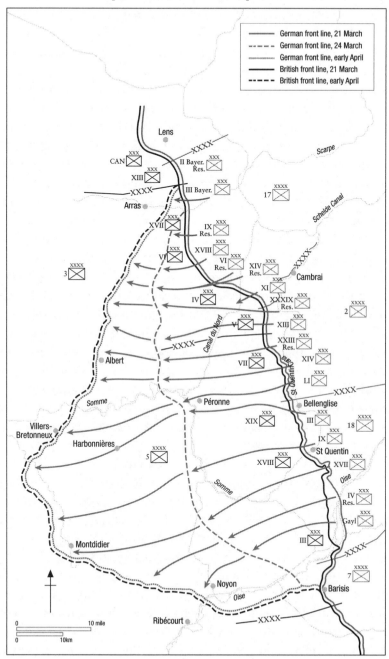

14 French divisions, prepared to launch a major counter-attack. At 4.25am on 18 July, without any preparatory bombardment, this huge counter-attack force threw itself at the exhausted Germans and began the Second Battle of the Marne. The attacking divisions advanced behind a massive creeping barrage towards ten depleted German divisions dug in behind very weak defensive positions. German machine gunners defended their positions gallantly but were systematically eliminated by the relentless advance of Allied tanks co-ordinated with dismounted infantry and artillery fire directed from aircraft above the battlefield. Worse than losing the Marne salient, the German High Command knew that they had lost the initiative. A general retreat to save what was left of that part of the German Army was ordered on 20 July. It was not a pretty sight. Discipline broke down in many units, looting was commonplace and the German war effort, so ordered and so efficient hitherto, was rapidly falling apart.

On the British front, the anxiety, fear and near-panic of March and April had been replaced by a determination to win the war, if possible in 1918. Fragile domestic politics at home, apart from anything else, made this highly desirable, but the rising cost in fatalities at the Front made it an imperative. The German failure to realise the operational significance of Amiens provided a major opportunity. It was from here that the British offensive of 1918 would be launched on 8 August and the countdown to the 'Hundred Days to Peace' would begin. The BEF was reinforced by divisions returning from the successful Sinai and Palestine campaigns and from Italy as the British involvement on that front was scaled back. Moreover, sensing the chance to win the war in 1918, and following a long and acrimonious struggle with Haig on this subject, Prime Minister David Lloyd George agreed to the release of reserve divisions previously kept back in England. Although Foch, who was promoted to Marshal of France on 6 August, remained as the overall Allied commander, Field Marshal Sir Douglas Haig gained Marshal Foch's agreement for the reinforced BEF to strike on the River Somme to the east of Amiens. It was known through intelligence from captured prisoners that the German Second Army, previously the tip of the German spear in March and April, was now weak and suffering with poor morale. Moreover, the ground in that part of Picardy where the attack would be mounted was highly suitable for the deployment of tanks, unlike the water-logged battlefield of Passchendaele the year before.

In the early morning of 8 August what became known as the Battle of Amiens began. Over 2,000 artillery pieces opened up at 4.20am, turning night into day and creating a hell on earth for the German defenders. It was a misty morning and the artillery barrage included copious amounts of smoke which added to the general confusion of the defenders. Their familiar practice of sending up signal flares to call for their own artillery support largely failed as the flares were often not seen, or when they were observed the German artillery batteries were already under intense counter-battery fire by the British gunners. The advancing British, Australian and Canadian divisions were supported by over 500 tanks whose presence added to the fear of the dazed German defenders. While some Germans fought stubbornly, others chose discretion over valour, surrendering at the first opportunity. The British Fourth Army's creeping barrage moved forward relentlessly on its pre-ordained timetable, followed by the tanks and the advancing infantry. Forty-five minutes after the British attack began, the French First Army to the south of General Sir Henry Rawlinson's Fourth Army began its attack too. Its objective was to take the town of Montdidier. By the end of the day not only had Montdidier been captured but a 15-mile gap had been opened up in the German line south of the River Somme. The potential for British exploitation deep towards the German rear areas added to the fear and confusion amongst the defenders.

General Erich Ludendorff called 8 August 'the Black Day of the German Army'. That day not only had a massive gap been created in the German line, but the Germans lost over 30,000 men, including more than 17,000 taken prisoner and 330 guns seized. Over the next three days the Allies continued the advance, driving the Germans back more than 12 miles. Gradually the momentum of the attack slackened as the Allies outran their own supplies and artillery support. Meanwhile on 10 August the Germans began to withdraw from the St Quentin salient created in the early days of their Operation *Michael* offensive in March and April. Ludendorff ordered a general withdrawal back to the Hindenburg Line where the German Spring Offensive had begun. Over the next six weeks the Allies kept up the pressure on the Germans, harrying them in a series of battles until the Hindenburg Line was reached.

On 26 September, Marshal Foch launched the co-ordinated Grand Offensive by the Allies against the Hindenburg Line. The first attacks were launched from the south of the Allied line by the French and Americans in the Meuse–Argonne area. On 28 September the Northern Army Group, consisting of the Belgian Army, Second British Army and French Sixth Army all under command of King Albert of Belgium, attacked near Ypres in Flanders in what became known as the Fifth Battle of Ypres. On 29 September, in the centre, the British Fourth Army and French First Army attacked in the direction of St Quentin. By 5 October the Allies had broken through the entire depth of the Hindenburg Line in the north and centre along a 19-mile front. On 8 October the British First and Third Armies continued the attack, breaking through the Hindenburg Line at the Second Battle of Cambrai. Reflecting on the recent fighting, General Sir Henry Rawlinson of the British Fourth Army remarked:

> Had the Boche not shown marked signs of deterioration during the past month, I should never have contemplated attacking the Hindenburg Line. Had it been defended by the Germans of two years ago, it would certainly have been impregnable ...[1]

This series of disasters forced the German High Command to accept that the war was lost and that an armistice must be negotiated. Unrest in Germany itself was widespread, the Kaiser was under pressure to abdicate and the morale of the army was very low as it was forced back through the territory captured in the heady days of 1914. The fighting continued in a further series of battles until 11am on 11 November when the guns at last fell silent.

In a supreme twist of irony, the leading elements of the British Army reached the Belgian city of Mons just before the Armistice was signed. It was in Mons that the first British soldier of the First World War, Private John Parr, lost his life on 21 August 1914 and where the last British soldier, Private George Ellison, was to die on 11 November 1918 just 90 minutes before the Armistice was signed. Both are buried within yards of each other in the military cemetery at St Symphorien just outside Mons.

Chapter 2

Confronting the enormity of the Great War on the front line

On the afternoon of 20 August 1918, Second Lieutenant Alfred 'Duff' Cooper, later to be Secretary of State for War and 1st Viscount Norwich, sat with the other officers of the 3rd Battalion of the Grenadier Guards at Saulty on the Somme. They had gathered to hear Lieutenant Colonel Andrew 'Bulgy' Thorne, their commanding officer, give his final instructions for the forthcoming battle.[*] To the young platoon commander it sounded like a 'hazardous enterprise' although he felt confident. The men of 10th Platoon were in good fettle. They had been well trained, the previous weeks being a busy whirl of preparation for the forthcoming battle to break the powerful Hindenburg Line.

After a hasty meal he and the men of his battalion were transported by trucks the 12 miles to Hendecourt-lès-Ransart, a few miles south of Arras. The objective of the first phase of the attack was the railway line running south from Arras to Albert. The Grenadier Guards were to secure a position on the railway line about three miles south-east at the village of Moyenneville. At Hendecourt they moved off to their assembly areas to await the 'off', which was to be 20 minutes after

[*]As a major general A.F.A.N. Thorne commanded the 48th (South Midland) Infantry Division in the defence of Dunkirk in 1940.

the artillery barrage began at 5am the following morning. Tea and biscuits awaited them, and as the night mist settled over the village the men completed final preparations, priming grenades, loading magazines for the Lewis guns and getting rid of all extraneous items from their webbing and packs. It would be left behind at the start point, to be looked after by the company quartermaster sergeant. The only kit they were to carry was ammunition (bags of Lewis gun magazines and grenades) and water. As they worked, Cooper chatted with his men. All, he noted, were cheerful. When there was nothing else to do, most fell asleep, their heads on their knapsacks or on sacks full of primed hand grenades – the ubiquitous No 36 'Mills bomb'. A few remained awake, talking quietly to each other over shared cigarettes and endless mugs of hot tea.

When at the designated time the barrage began, after several hours of quiet over the battlefield, it was the loudest he had ever heard. At 5.20am on the dot, without any whistles, bugles or shouted exhortations, the men arose from the ground and advanced into the darkness, going forward in groups around the two Lewis gun teams. The mist was so thick that Cooper's usual recourse – to follow the stars – didn't work, and the platoon very quickly got lost. In fact they strayed far to the south of their axis of advance. The battlefield was all confusion, but it didn't seem to matter. 10th Platoon, as they had been instructed, just kept going in what Cooper hoped to be the right direction. Every now and again they bumped into groups of men from other battalions. His diary recorded:

We met an officer in the Coldstream with a platoon. He said the Scots Guards had failed to get their objective, that everyone was lost and that the trench we were in was full of Germans. I said I would work down the trench, which I thought was Moyblain, our first objective, and clear it of the enemy. We went on, and presently I met Alec Robartes and Fryer.* The latter

---

*Captain the Hon A.G. Agar-Robartes MC, adjutant, and Captain E.R.M. Fryer MC, commanding No 1 Company.

was commanding No. 2 Company, which should have been in support of our No. 3 Company, but which had, although I did not know it at the time, already got in advance of it. After this I pressed on alone with my platoon, guiding myself roughly by the sound of our guns behind us. We were occasionally held up by machine-gun fire and we met one or two stray parties of Scots Guards without officers. Finally we met a fairly large party of the Shropshires, who I knew should be on our right.[*] The officer with them did not know where he was, but we agreed to go on together.[1]

Occasionally small groups of enemy were discovered in trenches and posts scattered across a deep battlefield. The concentrated Lewis gun fire usually quickly disposed of any potential opposition. When a trench needed to be taken the machine guns would attack from one side and grenadiers – men with especially strengthened Lee Enfield rifles with grenade attachments on their muzzles, capable of throwing a Mills bomb up to 200 yards – would attack from another. The assault sections – men with rifles and bags full of grenades – would only crawl up to the enemy trench under the cover of machine guns and the bombs fired by the grenadiers. Cooper found no difficulty in clearing enemy trenches. Each one that they attacked, fell. After an initial fight most enemy surrendered, the prisoners being sent back to the rear, before the platoon gathered itself together and continued on.

They knew they couldn't miss their objective, the railway line running south from Arras. They'd been practising on models carefully built in the ground for the past week. After a while Cooper realised that he'd drifted far to the right, bumping up to the village of Courcelles le Comte. He needed to move left. As he did so they encountered a machine gun firing straight down the road at them along the obvious route for him to take. The broken bodies of its

---

[*]These were men of the 7th Battalion King's Shropshire Light Infantry, advancing on the right of the Grenadier Guards.

earlier victims lay scattered on the road. Just then a lone Mark V tank trundled up through the early morning mist, also lost. When they called for its assistance, the tank advanced up the road and dealt with the machine gun, the men of 10th Platoon following on either side of the road through the fields. Reaching their objective, a building on the railway embankment, the Lewis guns made short work of the German defenders, those not being killed tumbling out with their hands raised.

So went Cooper's war. As Haig was to note in his diary, by 1918 it had been a platoon commander's war.[2] Despite the expected confusion of the battlefield Cooper did what he had been instructed: carry on forward in the direction of the objective, only fighting those enemy that directly imperilled his advance. The artillery barrage had opened the battle for them by targeting enemy machine-gun positions, command and control bunkers and artillery positions, as well as plastering the front trenches of the Hindenburg Line. The aim was to knock out the German defender's ability to strike back against the hundreds of dispersed platoons of British troops infiltrating through and behind the German positions. Cooper's men were told not to stay and fight for individual trenches if they didn't need to, but to keep pressing on; isolated enemy positions could be cleared up later by troops following behind. This was no longer a linear battlefield with everyone going forward in choreographed unison, but one in which semi-autonomous groups of machine-gun- and grenade-equipped infantry infiltrated their way forward to objectives far to the enemy's rear. The short bombardment was designed not to destroy the enemy trenches, but simply to provide the opportunity for the British assault platoons to get into and across the first German positions while their occupants were still down in their deep bunkers underground.

The wider battle in which Duff Cooper fought and distinguished himself – he was awarded the Distinguished Service Order for

his exploits in August 1918, an unusual award for a platoon
commander – was a very different one to that which remains firmly
lodged in Britain's corporate cultural memory of the Great War.
The shoulder-to-shoulder infantry advance that for most people
captures the horror of the Battle of the Somme on its inglorious
first day on 1 July 1916 had long disappeared. Battalions were
given objectives to secure and a considerable degree of latitude in
how to achieve them. In the case of Duff Cooper's battalion of the
Grenadier Guards on 20 August the plan entailed two forward
companies advancing towards the railway line under a creeping
barrage, supported by tanks. A heavy initial artillery barrage
would send the Germans scurrying for the safety of their deep
bunkers, there to await an attack, they thought, perhaps several
days later. On this occasion, however, the British assault platoons
would be in the enemy front-line trenches long before the enemy
were aware of their presence. Companies and their platoons were
expected to fight their way forward, independently if necessary,
relying on the initiative of their junior leaders, who would make
decisions based on their training and the circumstances of the
battlefield as they found it. Once the first two companies had
secured their positions, two further companies would pass
through each of them, carrying on the advance to the next bound.
During this advance they would be supported by the rifle and
machine-gun fire of the first two companies, who by now would
have dug themselves into hasty 'shell scrapes', prepared to see
off any attempt by the enemy at a counter-attack and to protect
themselves from incoming artillery fire.

Each of the ten platoons deployed for the attack (about 40 men
in each) was based on four sections, three of riflemen/grenadiers
and one with two of the portable gas-operated Lewis machine guns
introduced in 1916. The rifle sections were equipped with rifles,
hand grenades and dedicated rifles adapted with cup-dischargers
on their ends, which threw the Mills bomb on a seven-second
fuse up to 200 yards. Rifles were now subordinate to the hand
grenade and the cup-discharger. Rather than being riflemen in the
old sense, in which the bullet and bayonet were an infantryman's

standard weapon, the men were now effectively machine gunners or grenadiers, weighed down with hessian or canvas sacks full of grenades for hand throwing and firing from the cup-dischargers. In the machine-gun section rifles were secondary, self-defence weapons only, the men grouped around the two guns. Two gunners were responsible for firing the weapon, with two or three men supporting each gun by carrying, changing and re-filling its 47-round circular magazines. The old (pre-1916) concept of a platoon consisting of three or four equal sections of riflemen equipped with a Lee Enfield rife and bayonet had passed into history, at least on the Western Front. The 1918 version of the infantry platoon was one in which the primary weapons were the machine gun and grenade.*

A little more than two years before Duff Cooper and the men of 10th Platoon gathered to prepare for their assault on the Hindenburg Line, Private Donald Cameron of the Sheffield Pals, the name given to the 12th Battalion The York and Lancaster Regiment, went 'over the top' to advance against the village of Serre as part of the Third Army's diversionary attack on the Gommecourt salient. It was the first day of the Battle of the Somme, 1 July 1916. Cameron recounted his memories of this day to an interviewer in 1984:

> The first wave went over at 0720. They lay down about a hundred yards in front of our own barbed wire. Then the second wave

---

* The specialist sections in each platoon were reversed just before the Second World War in a new publication *Infantry Section Leading* in 1938 (building on previous iterations of *Section Leading*). Sections became rifle and bayonet again, each coalescing around a Bren gun. The purpose was to make the sections multi-purpose and multi-disciplinary but in so doing the essential offensive task of each, with the idea of a platoon attacking an enemy position with specialist and mutually supporting weapons from various directions, was lost.

went over, and lay down about thirty yards behind them. During this time, there was high explosives, shrapnel, everything you can imagine, coming over. Terrific hurtling death. It was soul destroying, but I wasn't frightened: I was impatient, I wanted to get moving.

The night before, they'd laid tapes, showing us the way to the cuts in the German wire. But when we went over, these tapes were missing, so we headed off in what we thought was the right direction. We'd been told that we had to walk at arm's length from each other, and that's how we started. But not for long. When we saw people dropping like ninepins on either side, we bent double, and in the end, we started crawling. After a while, three of us, and Sergeant Gallimore, got down into a shell-hole. It must have been about eight o'clock. The firing went on, and we kept peeping up, looking over the top to check, and the bloody Germans were sniping our wounded. They were even firing at the dead. They couldn't see us in our shell hole. I must have prayed a dozen times, I used to go to church when I was a lad, but I prayed more in that shell-hole than I ever prayed in a church.

We were dead tired. When it was dark, we found our way back to our own trenches, where there was a rollcall. Out of the eight hundred that went over, only a hundred answered. The rest were either wounded or killed. At the time, our parents used to send out food parcels, wrapped in cloth. So, there were parcels for eight hundred men waiting for us, to be shared amongst a hundred.[3]

This shocking slaughter remains largely how Britain remembers the First World War. It was a war the popular view recalls in which 888,246 British and Commonwealth soldiers went to their senseless deaths on a chaotic slaughter overseen by imbeciles. Thus the belief that the Great War was managed incompetently by elegantly uniformed châteaux-bound generals out of touch with the slaughter they were orchestrating became a widely uncontested opinion in Britain. The idea is brilliantly though

fictionally depicted in *Blackadder Goes Forth* and by Siegfried
Sassoon's 'The General', who 'did for' Harry and Jack 'by his plan
of attack'. Dunderheads there certainly were, but the bigger story
of 1916–18 was of soldiers and generals rapidly developing the
ideas necessary to transform warfare and bring about a victory on
the battlefield in 1918. The accusation of incompetence is closely
associated with the poets' claim – think of the battalions of writers
marching in Sassoon's footsteps like Laurence Housman, Leslie
Coulson, W.N. Ewer, Robert Graves, Wilfred Owen, Edmund
Blunden, Frederic Manning and others – that the war was bungled
by incompetent red tabs. (Most of these also believed that the war
was ultimately futile, as their diaries revealed.) The shocking losses
on 1 July 1916, the first day of the Battle of the Somme, in which
19,240 British and empire soldiers were killed, continues to fuel
this narrative. This unprecedented level of casualties – unknown
in these islands since the Civil War 300 years before – caused a
sense of shock that every subsequent generation has struggled to
comprehend. (By contrast, Europe had long experienced the high
casualties of mass warfare – indeed, the Red Cross was formed
following the high level of casualties incurred at the Battle of
Solferino in 1859 – and was less astonished at the mass destruction
entailed by industrial combat.)

The tragedy of the Great War was that it was a struggle for
mastery on the battlefield that took everyone at the time –
soldiers and politicians – by surprise. It was also hugely costly
in terms of blood, far beyond anything Britain had seen since
the self-inflicted follies of the Crimea six decades before.
The real problem for the tiny British Army in 1914 was the
incomparable scale of war's demands. No one in government or
senior command before 1914 ever anticipated the enormity of
the challenges placed on the country, and its army, during the
four years of war in France that followed August 1914. (This is
without considering the fighting elsewhere, such as at Gallipoli
in 1915, or the Mediterranean or Middle East in 1917 and
1918.) The developments in the technology of firepower over the
previous half-century coalesced with the raising and transport

of mass armies. This type of warfare eventually required the mobilisation of entire populations for the first time. Initially, neither side could dominate the other – whether with firepower, manpower or economic resources – such that victory could be quickly achieved. Instead, stalemate prevailed.

———

In 1914 Britain and its empire joined a coalition of nations – France, Belgium, Russia, with the addition in 1917 of the United States – in opposing the German invasion of France and Belgium. Following the encounter battles of 1914, in which the German attempt to encircle Paris failed, the war had stuttered to a stalemate along a 400-mile confrontation line stretching from the English Channel to Switzerland. The Germans, unable to press their offensive ambitions, now stood on the defensive along the River Aisne. In time Germany's strategic goal turned from crushing France in a six-week *Blitzkrieg* (to use this term before its time) to one that bled the country 'white' of its will and its physical capacity to resist. The German soldiers wired themselves in and dug themselves down. They built up trench positions of formidable strength, protected by yards of barbed wire, concrete strong points and machine guns. The Allies were given no choice but to go on the offensive to attempt to remove the invader.

These early battles established the format of the years that would follow. The challenge for all combatants was not just to break into an enemy position, for what use was it merely to secure the enemy's front-line trench? The purpose of a break *in* was to achieve a break *through*. This was the key that would then open the door to allow a break *out*, through which the cavalry would flood in a splurge of colour and glory and the battle would be won. The former might offer a limited tactical success; the latter would offer the possibility of a strategic victory. But to achieve break *out* required sufficient reserves close enough to the point of breakthrough to be deployed quickly (and before the enemy could deploy their own reserves) as Ludendorff discovered to his cost in March 1918. It needed commanders to be close enough to the front to be able to make this decision, and for adequate communications (telephone or radio, rather than human runners) to enable their orders to be disseminated with alacrity. Sequencing these three elements to bring about a strategic result was to be the ultimate goal of the wartime leaders. It was not to be achieved until 1918, first by the Germans in March and then finally, and decisively, by the British, Commonwealth and American armies in the Hundred Days battles that brought the war to a close.

With stalemate across the battlefield, the challenge was to understand how a successful breakthrough could be achieved. This was both a technical and a tactical problem. It was also a problem of scale. Men, guns and bullets were required in extraordinarily large numbers, together with the logistical paraphernalia needed to support armies that now numbered in their millions. The extent of the difficulties so presented had never before been contemplated. They entailed the entire mobilisation of Britain's domestic economy and its industrial complex, together with that of its empire. They required the development of new battlefield tactics and new technologies, such as aircraft, tanks and gas, to break the deadlock. They called for a massive increase in the quantity and quality of artillery, and the tactics used to make most use of this powerful weapon. In 1914 Britain manufactured 91 guns or howitzers; in 1916 it was 4,500; in 1917 it was 6,500; and in 1918 it was 10,680. Unprepared in 1914, Britain was only producing the quantities of artillery ammunition – at the requisite quality – for the tasks demanded of it, by 1918. All the while it had to work in close conjunction with its French allies, to ensure that the entire line remained strong at all points, and that stratagems for both defence and offence were co-ordinated. It had to calculate a means of breaking through the German line, with all the inherent advantages the enemy enjoyed of depth and solidity, and to do so – as its army grew – with newly recruited troops and newly promoted (and thus dangerously inexperienced) commanders. But it also had to wear down the fighting power of the German Army. This required a degree of attrition. Bludgeoning had a purpose, albeit until better strategies could be devised.

At any point in time the British Army is a product of many things: its higher direction; its people; its history and ethos; its fighting doctrine; and its equipment. Government policy is central to determining what tasks the army is called upon to undertake, against whom and for what purpose. It is the responsibility of the government to determine what Britain's security interests are, and to allocate resources to ensure that the country is prepared to deal with these threats as and when they arise. It is the task of the army to organise itself to meet these threats, to recruit and train its people, and to equip its forces within the resource constraints provided. To be effective the army needs therefore both to understand and reflect the grand strategic imperatives of the country as a whole, and to make sure that it has organisational and doctrinal – or 'warfighting'* – methodologies to meet these requirements. Where the country cannot identify a clearly stated strategic security interest, the army will struggle to structure or prepare itself meaningfully, as its ultimate purpose will remain unclear. Likewise, if the threat is clear and yet the army remains unprepared – structurally, doctrinally or in terms of training, morale, equipment or the consent of the nation – defeat in battle beckons. Herein lies the central dilemma for those who have responsibility for the nation's defence.

In addition to good doctrine, armies need coherent operational concepts for both defensive and offensive action, especially when they are locked into a coalition as the junior partner – a situation as familiar to British generals in France in 1914 and 1939, as in

---

*In this book we use the double-word 'warfighting' to describe being able to fight a peer adversary in high-intensity combat, deploying modern weapons and capabilities in an intelligent and co-ordinated manner, on the basis of well-defined doctrine and with well-resourced, well-trained forces led by highly competent commanders.

Iraq in 1991 and 2003. Therefore, they need flexible and adaptive commanders, able constantly to assess the war as it develops in order to refine tactics, equipment and approaches to battle. They need sufficient numbers of well-led, well-trained and well-equipped troops, able to co-ordinate all the features of combat power on the battlefield. They need weapons and technologies that are designed to enable tactical doctrine to be achieved on and above the battlefield.

For most of its history the British Army has been an exclusively volunteer force, so the support of the population from which its soldiers come has always been key to its ethos and motivation. Traditionally it has been corralled into local tribes – known as regiments – each with its own customs, uniforms and ways of doing things. This is because troops have always been raised locally, often directly by their own non-commissioned and commissioned officers. For all of their service, men remain in their regiment of enlistment, wear its particular marks, are commanded by officers they know and develop fierce loyalties to their comrades. The army's experience of fighting wars – what we describe in this book as the art and techniques of *warfighting* – have also been important, not least in defining how the army at any point in time thinks about how its future wars and campaigns are to be conducted. It's an easy – and usually unfair – gibe to accuse military commanders of trying to fight the last war. There is an imperative to analyse the past to produce a clear intellectual and practical appreciation of *how* war is fought, and the means available in the future whereby success in battle and in campaigns can be achieved. This a product of historic experience, contemporary awareness and analysis of the future. The mechanism for articulating this is doctrine.

The British Army had 247,432 troops at the start of the war, a third of whom were committed to imperial defence. When the British Expeditionary Force went to France in August 1914 it comprised 150,000 men in four infantry divisions (each of about

18,000 men, including 12,000 infantry and 4,000 gunners) a cavalry division (about 9,000 men), an independent brigade and support troops. It was a smaller deployment than even Belgium could manage. Britain's initial contribution was about 6 per cent of the total number of divisions in the Allied armies. It was very much the junior partner to France, which deployed 62 infantry and ten cavalry divisions. (The German invader deployed 100 infantry and 22 cavalry.) Britain would always be the tadpole in the pond, its strategic interests subservient to those of France. By the time General Douglas Haig came to command the BEF in December 1915 it had increased fourfold to some 600,000 men in 39 infantry divisions (including two Canadian). It comprised a mix of regulars, reservists and Territorials, with Kitchener's armies starting to arrive towards the end of the year. By the following year, one seared in British memory because it was the year of the Somme, the British Army as a whole had increased to 1.5 million all ranks, a sixfold increase in 18 months. Even so, the BEF remained the junior partner in terms of size, with 61 divisions in 1918 (51 British and ten empire) to about 200 French: 30 per cent of the overall effort. But for Britain these numbers were unparalleled. Its army had never been so huge. In the four years to 1918 it had enlisted 5,704,416 men, the army in the final year of the war standing at 3,820,000. By the end of the war the Royal Artillery alone exceeded the size of the British Army in 1914. At the start of the war it had 554 batteries; in 1918 there were 1,796. However, it wasn't just size that mattered, but how these troops were used on the battlefield to bring about victory. Could the British Army achieve this goal?

Chapter 3

Finding a way through the mud and the blood to the green fields beyond

Shockingly, by the end of the first set of encounter battles in 1914 the BEF had lost 90,000 men – a full 60 per cent of its deployed strength. This devastating savaging of the old pre-war regular army and the loss of such a vast reservoir of military experience was made up numerically by the part-timers of the Territorial Force and reservists – many long into civilian retirement – recalled to the Colours. In a realisation that this was to be a war of human quantity, the newly appointed Secretary of State for War, Field Marshal Lord Kitchener, began recruitment – amidst a groundswell of popular pro-war sentiment and goodwill – of a New Army to fill the gaps.

Unlike the professional and well-trained British Army in 1914, the new armies that confronted each other in 1915 and 1916 were armies of mass. They had a dynamic of their own. Men wanted to get the job done and get home. Tactics and approaches needed to be simple, as time for training and the garnering of experience was limited. In democratic countries, large numbers of intelligent, creative and motivated individuals join the ranks of their armies for the duration of a war. If these characteristics are not snuffed out by the sometimes overbearing pretensions of peacetime soldiering, the creativity of these men and women can be used to the army's considerable benefit. But first and foremost, soldiers need to be taught simple conformity to be able to operate together, with straightforward drills. By 1916 the army – the first mass army Britain had ever raised, from a country which had no tradition of conscription – was therefore a mixture of old and new. The old professional army (and its 'regular' reservists) had been joined by the Territorial Force, mobilised for war for the first time and sent abroad, also for the first time. Next came the men who volunteered at the outbreak of war or who had subsequently answered Kitchener's call for volunteers – the 'Your King and Country Needs You' campaign – in August 1914 and the months following. In 1916, their ranks were then swollen by the influx of the first conscripts. So, it was a large, heterogeneous and inexperienced force that went into battle on 1 July 1916. It inevitably made mistakes. Everything about the type of warfare encountered in France was new. No commander had had experience in the scale of command now presented; the technology to communicate effectively did not exist and nor did a tactical template for dealing with all the circumstances thrown up – in its intensity and scale – by this new type of fighting.

The first year and a half of the Great War saw the British Army attempting to understand how to retain offensive action in the face of a trench-based stalemate. Mid-1916, with the start of the three-month Somme offensive, saw the beginning of a concerted process of learning, about what worked and what did not, that was to culminate in the defeat of the German armies in late 1918. The war was a learning process, in which improvements – sometimes bold and sometimes faltering – were nevertheless gradually though relentlessly made. Technology improved (especially in terms of artillery). New ideas were conceived and adapted (such as the tank) or copied from the enemy (such as gas). More important than anything else, a clear understanding was arrived at as to how to use artillery to achieve breakthrough, enabling attacks to be sequenced accurately and methodically. The British Army's approach to warfighting – its tactics, command, logistics and much else – was adapted by trial and error over the period of the Great War. The Somme was the starting point of this learning curve, not the culmination of it, because it was only at the Somme that all the ingredients of the newly minted British Army met together for the first time.

One of the difficulties encountered in this process of learning was that in 1916 there was no single body responsible for collating learning, publishing the results and disseminating them across the army. Where it happened well, it did so within divisions, corps and armies, and within the service branches, not across the whole, but at the instigation and direction of their principal commanders. The war produced a mass of 'Notes from the Front', training pamphlets and instructions, but these all had a specific purpose and did not constitute army-sanctioned tactical doctrine. The army, as a whole, needed to learn how to learn. All too often the problem in the early years of the Great War was that doctrine represented the view of the most senior person present. The war had to be fought as it was encountered, problems had to be solved as they were faced, solutions identified and deployed in action to deal with the new exigencies of combat dominated by long-range artillery ammunition, machine guns, smoke and gas. It was fortunate that most senior commanders responsible for determining tactics were not the butchers of legend but intelligent and capable men grasping with – and in the main solving – the greatest military dilemmas of the age.

During 1915 the challenges for the British Army were formidable. How could it recruit a huge new army and deploy it successfully in battle? In France, how could the BEF not just survive trench warfare but become stronger and effective? The primary challenge for the British and French, given that their task was to eject the Germans from French and Belgian territory, was to learn how to launch and sustain successful local attacks on enemy positions. What techniques and technologies worked, or didn't? How could attacks be co-ordinated across the front to break the trench deadlock? The British attacked at Neuve Chapelle in March 1915 and Aubers Ridge in May, the Germans countering with gas for the first time at Ypres in April and May. The French tried in Champagne and both the British and the French launched a joint offensive at Loos on the first anniversary of the war in September, with combined casualties of over 96,000 men. The lesson was that even primitive defences held up against attack, as the Germans had found to their cost at Second Ypres in April and May 1915. The French counter-offensive in Champagne between December 1914 and March 1915 had experienced the same fate, with a new phenomenon savaging the high hopes and martial pretensions of the combatants: mass casualties. In Champagne, the attacking French lost 93,432 men to half that number of defending Germans. At Neuve Chapelle the British offensive, designed to conform with a French offensive (and to persuade the French that the BEF would and could fight), achieved some tactical success but at the cost of 12,500 men, about 30 per cent of the force engaged. Whatever else it was, this new type of industrialised warfare – and especially the offensive – was very expensive in blood.

The following year saw the first offensive undertaken by the newly enlarged British forces on the Somme, in support of the French resistance to German attempts to 'bleed the French Army white' at Verdun. Although the British planned and hoped to break through, the strategic purpose of the battle was simply to cause casualties to the Germans to prevent them reinforcing Verdun. Tanks were used for the first time at Cambrai the following year in support of a conventional infantry offensive.

In 1917 the biggest challenge to the Allies came with mutinies in the French Army in June, which left the bulk of the offensive effort falling on the British at Passchendaele, battles that were arguably in the wrong place and at the wrong time of year. As has already been outlined, the strategic gloom was finally lifted by the arrival of the US Expeditionary Force, although the March offensive by the German Army in the spring of 1918, despite being weakened by losses at Passchendaele, was designed to smash through the British and French lines at Amiens before the Americans were able to deploy offensively for the first time. At the time, the planning challenge for 1918 and 1919, and for as many years after that the war ran on, was to launch an effective series of offensives to exploit the German failure in March, to turn the tide and bring the war to a close. In mid-1918 that prospect looked far off. What was to be done? Success would require a thorough analysis of warfare as it had been experienced so far on the Western Front, together with careful thought and application as to how it could be undertaken differently in the future.

In 1914 the British Army's doctrinal theory was contained in the Field Service Regulations, or FSR, first published in 1909. Part 1 – 'Operations' – was a 296-page summation of the military wisdom required to manage an effective campaign. Written by Brigadier Generals Henry Wilson and Henry Rawlinson (not, as some have imagined, Douglas Haig), it captured the principles of successful campaigning against both a peer and an (in the language of the time) 'uncivilised' adversary. It concerned itself with principles and approaches, not with detailed tactics or drill, although in many places clear advice from wars of the recent past was provided. These examples were most notably sourced from the Second Boer War, but included wars in which Britain observed rather than took part, such as the Russo-Japanese War of 1904–05 and the Franco-Prussian War of 1870.

There is little in these pages that would surprise a modern soldier. For example, it is clear from the Field Service Regulations that the British Army understood firmly the decisive role played by artillery in the attack:

> The principle of the employment of artillery in the battle is that the greater the difficulties of the infantry, the more fully should the fire power of the artillery be developed. As the infantry advances to the decisive attack, every effort should be made to bring a converging artillery fire to bear on its immediate objective, and artillery fire will be continued until it is impossible for the artillery to distinguish between its own and the enemy's

infantry. The danger from shells bursting short is more than compensated for by the support afforded, if fire is maintained to the last moment; but in order to reduce this danger, it is the duty of artillery commanders to keep themselves informed as to the progress of their infantry, and to discontinue fire against the objective of the assault when the infantry is getting to close quarters if such fire cannot be readily observed and controlled. A portion of the artillery must be pushed forward so as to be able to deal with possible counter-attacks, and to give the infantry immediate assistance, when the fluctuations of the fight make this necessary.

The climax of the infantry attack is the assault, which is made possible by superiority of fire. The fact that superiority of fire has been obtained will usually be first observed from the firing line; it will be known by the weakening of the enemy's fire, and perhaps by the movements of individuals or groups of men from the enemy's position towards the rear.[1]

The problem lay in determining *how* an attack could be successfully achieved in the context of a gridlock of opposing, trench-bound armies, where the Germans had adopted a deliberately defensive posture and their opponent had no choice but to learn how to attack effectively and by doing so somehow bring about a strategic result. The challenge of the Western Front lay in finding a way to translate an admirable theory, written in a previous era of warfare, into effective tactics. The problem at the outset of the Great War was that the doctrine drafted so ably five years previously could not conceive of the mass movements of armies, the ability of devastating firepower to force troops underground, the emergence of new technologies (for example gas, aircraft, radio communications and light, infantry-borne machine guns, such as the Lewis gun) and the impasse caused by trench warfare.

When the Field Service Regulations talked about artillery support to the infantry during the attack it had in mind the wheeled guns of the Royal Field and Royal Horse Artillery. By 1914 each infantry division of the BEF took with it 54 of the excellent 18-pounder breechloading, rifled guns.* As a result of experience in the Second Boer War their primary ammunition was shrapnel. The Territorial Force had to put up with the older 15-pounders. (Divisions also had 18 4.5-inch howitzers and four 60-pounders firing high explosive shells, designed for attacking hardened positions.) Indeed, in 1914 large numbers of casualties – French, German and British – were caused by artillery against unprotected infantry. This led to the imperative to dig, recalling a lesson of the American Civil War half a century before. Trench warfare added considerable complexity to the artillery task. Likewise, the sheer amount of ordnance being thrown about on the 1914 and 1915 battlefield made many of the pre-war assumptions about offensive warfare sound naive.

Operations during 1915 further reinforced the importance of the guns in the trench environment. The battles fought by the BEF in 1915 – at Neuve Chapelle (March), Ypres (April/May), Aubers Ridge (May), and Loos (September–October) – demonstrated that infantry could often overcome enemy trenches if these positions were first overpowered by artillery. Artillery could do several things. High explosive could physically destroy enemy trenches, although the further down the enemy dug, the more protected they became. Both high explosive and shrapnel could keep the enemy's heads down during the infantry attack, so in theory assaulting troops could cross contested ground (so-called 'no-man's land') and occupy enemy trenches before the enemy were able to climb out of their dugouts. If it were accurate enough, it could destroy the enemy's own artillery, to prevent counter-battery fire during and immediately following an assault. High explosive could also, if their locations could be identified, destroy enemy strong points, such as defensive positions, command posts, machine-gun posts

---

*In 1915 the requirement to equip the New Armies and the Territorial Force had forced a reduction to 48 guns per division. In 1916 a further reduction to 36 allowed the formation of army field artillery brigades, a centralisation of assets which could then be deployed where the army (rather than divisional) commander thought they were most needed.

and such like. In order to be able to undertake these tasks four things were needed.

First, many guns, with much ammunition. In terms of guns, the army had far too few to be able to do everything commanders wished of them. What the onset of trench warfare demonstrated was the need for high explosive shells of heavy calibre. In terms of ammunition, the pre-war British Army doctrine was heavily weighted towards shrapnel. This was useful in killing enemy men and horses in the open. It had limited utility against trenches. Nor could it cut barbed wire, despite the belief derived from Neuve Chapelle in 1915 that it could do so (shrapnel dispersed the wooden apparatus holding the wire together, rather than cutting it). There was no high explosive shell for the 18-pounder at the start of the war. Its production required a significant reordering of industrial priorities in 1915, a process that cost some political scalps, including that of Asquith, the prime minister.* It was to take time to turn this situation around. Even by mid-1916 there simply were not enough guns or enough ammunition, especially of high explosive, to meet all the requirements of the army in France. In other words, the problem was as much a political one for Britain (making the correct procurement choices early enough) and an industrial one (providing sufficient manufacturing space to build the quantity of shells required), as it was for the general in the field deciding how to build artillery into his plan for battle. During the Battle of the Somme over 66 per cent of British shells fired were still shrapnel, because this was what the army had in its stocks. Many other problems required solutions. Constant use of artillery wore out gun barrels. The quality of shells remained poor, perhaps as many as 30 per cent on the Somme failing to detonate at all. Furthermore, cutting wire by the use of artillery remained

---

*The so-called 'Shell Scandal' of 1915. See Richard Holmes, *The Little Field Marshal: A Life of Sir John French* (London: Weidenfeld & Nicolson, 2004) and Hew Strachan, *The First World War: To Arms*, Volume I (Oxford: Oxford University Press, 2001).

an impossible task until the introduction of the instantaneous percussion fuse which splintered on impact rather than cratering.

The second requirement was for guns that could fire directly and accurately onto their targets, some of which were directly forward of the assaulting troops while others might be widely dispersed throughout the enemy's defensive area (second- and third-line trenches, specific strong points and so on).

Thirdly, sophisticated attack plans were needed that allowed a range of guns of different types and calibres to fire on distinct targets.

Finally, there was a requirement for artillery accurate enough to undertake precise attacks on identified targets, rather than to act merely as area bludgeoning weapons.

A plenitude of artillery, well-supplied with sufficient stocks of dependable ammunition and intelligently used across a variety of operational scenarios, while using every technological advantage discovered over the recent years of combat in France, was the new *sine qua non* of modern warfare.

Formed in January 1916, Rawlinson's new Fourth Army was the main British attacking force for the Somme offensive, comprising 11 divisions and nearly half a million men. It was a polyglot group of battered though experienced regular army battalions (each brought up to strength by reservists and new recruits) together with Territorial Force units which, though eager, had little experience. They were joined by the newly raised units of Kitchener's army, equally inexperienced. In 1916 the BEF in France was essentially a new army, with a mix of experienced troops and large numbers of men and officers who were eager though ignorant. To compensate for this inexperience an unprecedented artillery bombardment of the German first-, second- and third-line trenches would, it was hoped, destroy the German position entirely.

If 1915 demonstrated the art of the possible with artillery, 1916 showed up its limitations. One of the problems commanders had at this stage in the war – mid-way through the second full year of fighting – lay in understanding how best to manage the relationship between attacking artillery and assaulting infantry, enabling the guns to reduce infantry casualties. In the period between 1 January 1915 and the start of the Battle of the Somme, the number of artillery batteries in the BEF had increased from 36 to 191. Accordingly, Rawlinson's plan for the Somme was designed to save the lives of his infantry by letting the artillery do all the heavy lifting. The enemy trenches would be subdued by weight of fire; the wire obstacles cut by the blast of fragmentation. The infantry would need merely to occupy the devastated enemy trenches once the bombardment had ended. Far from being one of the butchers of modern perception, Rawlinson was actively seeking ways in which firepower could relieve the pressure on manpower. The attack plan for the Somme was meant in fact to save lives, not to throw them away.

Rawlinson believed that in addition to this bombardment by high explosive, the advancing infantry needed to be covered by protective fire from the division's 18-pounders, firing shrapnel on the enemy trenches to keep their heads down, or directly in front of the advancing troops, as the men crossed no-man's land, walking ahead of the men (a 'creeping barrage'). This was something first tested at the Battle of Loos the previous year, the artillery barrage jumping ahead of the infantry in carefully calculated synchronicity.

On paper, the details of the bombardment were astounding. The assaulting divisions were to be supported directly by 808 of their own 18-pounders, firing shrapnel in support of the assault, together with 202 4.5-inch howitzers firing high explosive. Additional French guns were loaned to Fourth Army to help. The job of the 18-pounders and their shrapnel was to cut through the wire in front of the German positions, and to fire the infantry onto their objectives. A further 427 heavy guns and howitzers would bombard other targets of interest. These would target enemy gun positions, ammunition stocks, machine-gun nests and headquarters positions. In total 1.5 million shells would be fired. It's hardly surprising that, in the circumstances, some British officers suggested to their men that the result would be a walkover.

In the event, the firepower failed. Guns had to register on their targets, alerting the Germans to British intentions days before the infantry assault and thus losing any element of surprise. The Germans made full use of their deep bunkers to survive this storm of fire. The problem was that in 1916 Fourth Army simply wasn't mature or competent enough to make this type of complicated artillery plan work. It was a ferociously complex task to calculate a fire plan based on the advance rate of infantry in a particular area of ground. It also used up vast quantities of ammunition, with each gun firing three or four rounds per minute. In some divisions the artillery coverage proved too shallow to be effective. In other cases, the gunners misread their instructions, and simply pummelled the front line, with the result that the infantry were left exposed in no-man's land when the barrage ceased or moved on. In other instances, the timings of the progressive creep forward were unrealistic, moving too fast for the trudging infantry, trying to make their way across a muddy shell scape, burdened down with equipment.

Equally, the operational inexperience of both soldiers and commanders on 1 July 1916 – infantry and artillery – showed through. Such a complex plan required preparation and rehearsal, little time for which was available. With so much else needing to be done – such as building roads, stores dumps and a million other things – the actual rehearsal for the assault fell down the priority list. The old Field Service Regulation method by which infantry rushed against an enemy position, a platoon lying down to give covering fire to another, before swapping the process and running forward in leaps and bounds towards an objective (known to the Army as 'fire and movement'), was disregarded for this attack. Why? Because it was considered too complex to be learned by the hundreds of thousands of new recruits in the time available and was impracticable to co-ordinate with supporting artillery. With such an enormous bombardment planned, the idea that the infantry would even need to rush enemy positions using traditional assault tactics was dismissed as too pessimistic by many commanders. (On several occasions during the battle this traditional tactical approach was used by some commanders, with good results.) In terms of the artillery, many of the guns – some of Boer War vintage – were by now worn out, and much of the ammunition defective. A further problem was that unless artillery observation officers could see the fall of shot, an advantage only on the higher ground of the southern sector, where all the successes of the first day were recorded, it was impossible to adjust the artillery fire to match the changing situation on the ground. Where were the accompanying infantry? Where were the enemy positions that required targeting? It was hoped that aircraft could take photographs and feed this information back to the gun lines. Fog in many places of the front on 1 July 1916 made this impossible, and, in any case, getting photographs back could take as long as 24 hours. A solution to enable the artillery to follow the battle, rather than be stuck with a timetable that was worthless once the attack had begun, had not yet been devised. What were known as 'on call' fire plans, where guns were made available to attack targets when they arose, had not been perfected in 1916.

Given the anticipated effect of the guns for the Somme battle, Haig ordered Rawlinson to use his 202 howitzers to spread their fire across the second- and third-line trenches as well, disrupting all parts of the German defensive position along the Fourth Army front rather than merely the front line. The idea was a good one: it was not sufficient merely to defeat the front line when a second and reserve line of defences, not to mention enemy artillery batteries, remained to be subdued. The problem was that artillery was being asked to do things that it simply could not do. There wasn't enough of it in the first place, and the plan meant that Rawlinson's heavy guns, rather than concentrating exclusively on the front line during the preparatory bombardment, now had to deal with a far greater range of targets than he had first envisaged. The result was that Rawlinson's artillery effort was now spread much more thinly, reducing even further the chances he had of destroying the German front-line trenches, let alone all the others.

Haig wanted him to undertake a rapid or 'hurricane' attack of five or six hours, though Rawlinson refused. He did not have enough guns to fire the quantity of ammunition he considered necessary to cut the barbed wire in front of the enemy front line and demolish the German positions. Instead, he opted for a bombardment lasting five days ending just prior to the order for the infantry to climb out of their trenches and advance on the enemy. Who in their right mind, as the thinking of the day went, would have suggested before such an extraordinary bombardment that it would not have been possible for the infantry subsequently to cross no-man's land unhindered? If the assumption was made that such a bombardment would entirely destroy the front-line German defences and disrupt the enemy's second and third line, the requirement to protect the advancing infantry by using smoke (proven at Loos) or of using fast rushes (or columns) to get them out of harm's way quickly, were obviated.

One of the problems with planning for the Somme offensive in 1916 was that commanders had widely varying views of what could and could not be achieved by artillery; another product of the absence of agreed army doctrine. The pre-battle assumption was that the weight of the planned bombardment would successfully destroy or at least disable the enemy trenches. What happened if, however, when the infantry went over the top, the enemy trenches had not been destroyed? To be effective, high explosive was required for the former task, and plenty of it, although even it could not hope to penetrate into the deep living quarters the Germans now routinely built far below the surface. Effectiveness was a function of how much high explosive could be dropped on a position, and how accurately.

Haig, Rawlinson and their corps commanders in 1916 all assumed they understood the decisive role that artillery would play in the forthcoming battle, basing their judgements on personal experience of Neuve Chapelle and Loos. As well as fatally overestimating the power of their own guns to destroy the front-line trenches, cut the wire in front of them and neutralise the German artillery batteries, they had not accounted for the rapidly improving complexity of the

German positions, the deepness of the living quarters they were able to build far underneath the forward positions or the interlocking complexity of their machine guns. Nor had they accounted for the general inaccuracy of guns at the time, or for the wear of barrels, or for failures of fuses. In addition, in the months that had followed the great battles of 1915 little of this personal experience had been distilled into a slowly developing army-wide doctrine, defining the best way to fight an offensive battle in the new circumstances of the Western Front.

As a result, the plans for 1 July 1916 were based on a set of assumptions that were to prove to be wrong. First, that artillery on its own – or the type and quantity available on 1 July 1916 – could comprehensively destroy an enemy position. The Germans were fast mastering the art of digging down deep. Secondly, that the firepower needed to precede movement forward of the infantry, or be synchronised with it, rather than *accompanying* it, was the guiding idea of combined fire *and* movement.

In fact, the exact nature of each divisional assault was left up to its respective commander. Some troops in III Corps advanced in waves at the walk or at a steady pace, while some battalions in VIII Corps rushed the enemy's lines in waves. There was, as yet, no agreed or doctrinally determined way of overcoming an enemy defensive position, and certainly not one yet that did anything other than attempt to synchronise the infantry assault with its preceding *or* accompanying artillery bombardment. Where divisional commanders were reluctant to believe that artillery would work as promised, the results were very different. The divisional commander of the 36th (Ulster) Division insisted that his men advance close to the German front line during the artillery bombardment, rather than after it, rushing the enemy trenches just as the gun fire lifted and before the defenders could come up from their deep dugouts. As a result, they were able to secure the positions with minimal casualties. The general failure of the artillery to overwhelm the defences meant that Rawlinson's attacks failed at great cost.

The assumption that the infantry did not need to be accompanied by fire *during* their advance to contact, because the artillery bombardment that preceded it would be sufficient for this task, was an expensive lesson of 1916. The widespread use of smoke to cover the advance, direct fire by field artillery firing shrapnel ahead of the advancing troops, and the use of assault weapons (infantry-held machine guns, rifle grenades and trench mortars) among the first groups of advancing infantry would have dramatically improved the chances of success on 1 July 1916. But none was deemed necessary. Within a few days of the battle starting, however, it was recognised that a creeping barrage of shrapnel just ahead of the advancing infantry from the divisional 18-pounders was necessary to provide protection from enemy defensive machine-gun fire at the point when the advancing troops were at their most vulnerable. The ideal situation was where a creeping barrage would enable the advancing infantry to fall on the enemy's front-line trenches unmolested.

But perhaps the greatest unlearned lesson of 1916 lay at the operational level of war, rather than the prosaic (though bloody) issues of crossing contested ground. The rationale for the Somme offensive was to kill Germans, not to break through the German front lines. The battle was deliberately envisaged as a means to keep the Germans occupied in bloody slaughter while to the south the French were engaged at Verdun, thus preventing the Germans from reinforcing this sector with troops from the Somme. It was for this reason that Rawlinson determined the need to launch an offensive in equal parts across every inch of the Fourth Army's front line, rather than to focus his efforts on penetrating one part of the line. It was the reason why his initial plan was to concentrate his effort on attacking and occupying only the front line of the enemy trenches. Haig, who had replaced Sir John French six months before as commander-in-chief of the BEF, insisted that he attack more deeply, to include both second- and third-line trenches in order to have a chance of breakthrough.

The bitter experiences of failure across many places on the British front on 1 July 1916 were nevertheless not wasted. Despite the horrific loss of life, in part a consequence of the vast numbers of troops involved, lessons were learned quickly. For the next stage of the offensive, which began on 14 July, a raft of new features accompanied the plans, all of which had been identified from the experience on the first day of battle. Rawlinson exploited the success in the south around Montauban and launched a dawn attack on the villages of Bazentin-le-Grand, Bazentin-le-Petit and Longueval. Night attacks are notoriously complex to mount, as they require considerable degrees of low-level initiative to manage and to recover if things go wrong. But the advantages – of surprise and of being hidden from view to the defender's machine guns, not least – are significant. Instead of a long bombardment, the targets were placed under a short attack exclusively of high explosive lasting only five minutes, designed to force the enemy to get their heads down and to think that they were in for a long bombardment. Where infantry did advance under cover of their guns, they did so behind a creeping barrage of high explosive. In some places 'saps', shallow trenches dug at 90 degrees to the firing lines, in order to allow a jumping off point closer to the enemy positions than their front-line trenches, were dug at night, the soil disposed of to the rear to avoid alerting the enemy to their presence.

These attacks succeeded spectacularly, with a fraction of the casualties suffered in earlier engagements. Even inexperienced troops did well, so long as they were briefed thoroughly in advance, and had the opportunity to undertake dry runs on models built to enable all soldiers to understand the role they would play in the attack. The lesson of 14 July 1916 was that the British Army, led in this instance by Rawlinson's Fourth Army, was learning fast, almost day by day. This culture of learning, often from mistakes and bloody reverses, was a critical feature of the BEF from this point on. Much more needed to be understood, not least of all how to repeat the success of an attack into a second, third and repeated attack so that offensive momentum could be maintained.

Chapter 4

Masters of the battlefield, 1918

Despite the enormous figures – both for casualties and for material expenditure – which have become synonymous with the First World War, tactics and approaches to mitigate the shocking power of materiel in the offensive developed as the war developed. Artillery was a prime example of dramatic changes in technology and tactics.

Following the Somme, the employment of artillery fire became increasingly more sophisticated and flexible. More reliable fuses were in use, guns were matched with targets and accurate indirect fire began to be laid without direct observation. Predicted firing (from map and topographical survey) was increasingly employed, enabling guns to hide their positions before battle instead of giving away gun locations and potential targets to the enemy by the methodical registering of targets. Enemy batteries were sometimes identified by means of sound ranging, microphones collecting the distinctive low-frequency firing signature of enemy guns, and trigonometrical calculations pinpointing their exact location. Flash spotting, by which observers saw the distinctive muzzle flash of enemy guns, and plotted them on maps, was also employed. All these innovations meant that the British had accurate intelligence about most enemy artillery locations before battle commenced. When the time came, they could attack them without warning, bringing surprise back to the battlefield. In most cases, preliminary bombardments prior to an offensive were no longer required. In addition, by 1918 the Royal Artillery had instituted new mechanisms for collecting, collating and disseminating innovation and new ideas. Pamphlets were published and updated to ensure that every gunner was aware of the latest information about artillery matters.

The co-ordination of artillery fire, of all types and natures of ammunition, was now managed in a single *fire plan*. This was commanded, at divisional level, by the Commander Royal Artillery, a major general. Instead of the massive pounding of enemy positions which had characterised the earlier battles, attacks were increasingly nuanced. At Vimy Ridge in 1917 the fire plan was built on the principle of selective bombardment, rather than of destruction. Particular trenches across the enemy's defensive position were attacked. During bombardments the Germans often thinned out their front-line trenches, with the men being moved back to safer positions in the second line. For this reason, second-line, reserve and communication trenches were also struck, details being garnered by means such as map survey, sound ranging and aerial photography. Strong points and known machine-gun positions were also targeted, especially those sitting in defilade positions to potential avenues of attack, together with fuel and ammunition dumps, artillery batteries, headquarters locations and road junctions. At night, the attacks would continue on high-value targets. Aerial observation had by this time become an established part of fire control.

Improvements in artillery technology and practice were paralleled by continuous improvements in offensive tactics. General Herbert Plumer introduced innovative 'bite and hold' tactics at the Third Battle of Ypres in September and October 1917, in which objectives were set for advancing troops based on what the commanders were confident could be held against enemy counter-attacks. Tanks were developed and first deployed in 1916 to enable the infantry to cross the battlefield in comparative safety, and so counteract the effect of the machine gun. Combined-arms groups of infantry and tanks, supported by dedicated artillery and aerial support, were pitched at identified weak spots on the enemy's front, and fought as a coherent team to break through the enemy's line and continue in a deep penetration. The idea behind this type of attack was partly psychological. The earlier assessment was that a whole area of trench line would need to be captured, rather than one part, because otherwise the defenders would merely need to close ranks to cut off the intruders. But the psychological value of a thrust deep into the enemy at a single point or multiple points (rather than across a wide area of front) would often mean that defenders to right and left would be discomfited by the attack and fall back to reconstitute the line. What they should have done, of course, was immediately counter-attack to break up the enemy attack. This is hard to do, however, and takes considerable training, presence of mind, leadership and not a little courage. The psychological discombobulation caused by a successful thrust was such that, if it were repeated at other points in the line, a general withdrawal by the defenders might be initiated. If this happened, a managed withdrawal was often very difficult to achieve, such that the advantage would lie with the attacker. The challenge lay in keeping the attack going. Momentum is key. Reserves needed to be brought up, along with mobile artillery and the logistics required to fuel the advance.

These tactics emphasised speed, surprise, all-arms co-operation, and assault by self-contained small units, what the Germans were to call 'storm troops'. The Germans first adopted the assault troop principle in 1917 (although the French claimed to have devised the principle first) with a basic unit of 11 men, although aspects of this development can be found across both German and Allied armies from late 1916 onwards. Short artillery attacks (sometimes called a 'hurricane bombardment') on a narrow piece of front would be followed by infantry platoons and companies armed with machine guns and grenades, accompanied by tanks (in 1918) and artillery forward observation officers, able to call up the fire of guns in the rear to attack opportunity targets as they appeared. The storm or shock troops would be tasked with breaking through an enemy position across a narrow front, and then pushing troops deep into the enemy defences. This 'infiltration' would be followed up by more heavily armed troops with the task of destroying any fixed positions of resistance. The vanguard or infiltration troops would bypass areas of resistance to be mopped up by more heavily armed troops flowing up from the rear. Reserves would be brought up to help exploit what the opinionated military commentator Captain Basil Liddell Hart, a correspondent of the *Daily Telegraph* (1925– 34) and *The Times* (1934–39) would describe as an 'expanding torrent'.[1] The idea here was to exploit the effect of psychological discomfiture. If the defenders of a strong point knew that the attacking enemy had advanced far behind them, and that they were effectively cut off from resupply or reinforcement, they would perhaps be less eager to fight on, inclined instead perhaps to disengage and retreat.

The only way to counter these tactics was defence in depth. This idea was that front lines were actually elastic and focused not on trench lines but on fixed defences or bastions across a 'battle zone' often several miles in depth. Defence in depth is a stratagem to degrade the power and effect of an enemy offensive. It was designed to exhaust the attacker, and to break up an attack so that it was dispersed over the battlefield, constantly under threat from the positions it had bypassed but hadn't subdued, all the while becoming progressively weaker. An army needs confidence to fight in depth, knowing that the enemy is all around it. It requires confidence – the product of good training and leadership – to remain firm, and to counter-attack the enemy's vulnerable flanks and rear, rather than fleeing in panic. It takes confidence to persuade the enemy that it is they, not you, who are surrounded and cut off from their supplies and rear. But it also takes organisation to keep those posts cut off by the enemy advance supplied with supplies of food, ammunition, fuel and water, and it requires tactical skill, ingenuity and communication to support surrounded posts by counter-attacks and operations designed to reduce the enemy's freedom of movement, and to enhance one's own. And, overall, it requires calm and confident leadership to bring all these factors together so as to dominate the entire battle space. Enemy reserves need to be prevented from coming up, artillery needs to be able to interdict flexibly a wide variety of pre-registered targets on call, and the capability is required to resupply cut-off garrisons and posts.

The problem with mounting an offensive designed to break through an enemy position lay in sustaining each phase of the attack so that offensive momentum could be maintained. The objective was to break through the enemy defences, a task that required multiple consecutive phases of attack. The difficulty was that in each new phase the guns needed to be brought up and registered on new targets; logistical resupply had to be secured and fresh troops brought up and prepared. Defence in depth was designed to draw in, disperse and exhaust an attacker. The attacker therefore needed a stratagem for sustaining their attack. In March 1918, Ludendorff's offensive made initially impressive gains (relative to previous years of battle), although these were achieved at the expense of the German infantry. Once their men had stopped from exhaustion, there were no reserves to follow them up. The Germans proved able to initiate an attack, but not to sustain it.

The British and French forces caught on to the concept of defence in depth, albeit fitfully, such tactics leading to the failure, for example, of Ludendorff's 'Mars' attack launched against Arras on 28 March 1918, a week after the start of Operation *Michael*. Thirteen German divisions attacked six British divisions defending well-dug and well-wired positions with determination and skill. The British had withdrawn the bulk of the defenders far behind their front lines, with the result that the opening artillery bombardment fell largely on unoccupied positions. Defence in depth allowed the enemy attack to be absorbed. Minefields to channel attacking troops into artillery and machine-gun killing zones and anti-tank ambushes took their toll of the attackers, until counter-attack forces could be launched to mop up the exhausted enemy infantry.

By June 1918 it was clear to the men of Rawlinson's Fourth Army (one Australian corps, three British and one Canadian corps) that their opponents had lost the vigour they had possessed during the March offensive. Large-scale raids by the Australians discovered that the German defences were poor, the fighting quality of the men was not at the usual high standard and morale was low. Moreover, 'Spanish flu' had begun to weaken the German Army. Rawlinson and others began planning a devastating counter-offensive to make the most of the German discomfiture. By this time in the war Rawlinson's biggest problem was lack of trained manpower. Battalions were now down to around 650 men from the 1,000 they had comprised in 1916, and British divisions now had only nine rather than the earlier 12 battalions. This meant that a British division boasted 6,500 men compared to the 12,000 of 18 months earlier. There were no men left of military age to replace casualties. The mathematics of attrition were of inexorable decline.

The first solution to the problem was to look to the newly arrived American Expeditionary Force. A second was to rely on the massive output of Britain's heavy industry. Britain was now firing on all its industrial cylinders. The country was mass producing tanks, machine guns, artillery and artillery shells at a prodigious rate. Rawlinson recognised that this gave him the opportunity to use tactics based on firepower rather than manpower. In June he proposed improving the offensive power of his divisions by substantially increasing the numbers of Lewis guns per battalion to 64 (from 20) – an enormous four per platoon – and allocating 36 tanks to each. The role of the tank remained to support the infantry in the assault.

The first opportunity to test these developments was the Battle of Hamel on 4 July 1918. Lieutenant General John Monash, the newly appointed Australian Corps commander, designed a plan for Rawlinson based on surprise. The 600 guns allocated to the bombardment were all to be registered on their targets silently, so as not to give the game away. There were several new, innovative features of Monash's plan. First, 60 of the new Mark V tanks of the British 5th Tank Brigade would lead the assault. At 300 yards behind them would follow the two Australian brigades (eight battalions), whose target was not the front-line trenches, but rather those 2,000 yards to the rear. The infantry were equipped with additional Lewis guns and rifle grenades: their job was no longer to assault enemy trenches with rifles and bayonets, but to cover their own advance with machine guns and use rifle grenades. In another innovation, the advancing armour and infantry would be protected by a creeping barrage while howitzers fired high explosive against likely centres of resistance behind the German front. The enemy machine guns, long the *bête noire* of attacking infantry, would be subdued by a mixture of a bombardment from the heavy guns, the field artillery barrages firing shrapnel, and finally an assault delivered by tanks and infantry acting in concert. When the two forward brigades secured their objectives, the two fresh reserve brigades would advance and repeat the procedure. The use of the tanks enabled Monash to double the frontage of the attack, which spread the available infantry across a far wider area.

Several battalions of Americans joined the assaulting forces. Two-thirds of the 302 available heavy guns blanketed the German batteries as the infantry and tanks advanced, removing for several hours the threat of hostile artillery. With a couple of exceptions, the creeping barrage by the 18-pounders firing just ahead of the advancing tanks and infantry worked well. On the occasion where old-style frontal assault was undertaken without close support of artillery in its various forms, considerable casualties were incurred. Where tanks did not get forward, German strong points were overcome by infantry platoons constructed around Lewis guns with riflemen equipped with rifle grenades and trench mortars. It was the arrival of the Lewis gun, able to be handled by one man with another carrying the ammunition, which enabled platoon tactics and structures to be transformed. The transformation from 1917 was profound. Instead of a 30-man platoon with .303 Lee Enfield rifles in three or four equal rifle sections, the infantry platoon was now conceived as one which revolved around the Lewis gun, of which there were initially two per platoon increasing to as many as four in 1918. Light machine guns, rifle grenades, grenades and light mortars now constituted the offensive and defensive firepower of the infantry platoon. The rifle and bayonet still had utility, of course, but by far and away the most important weapons in this new type of warfare were machine guns and rifle grenades.

Hamel was a dramatic success. For 1,000 casualties, a fraction of those that would have been expected in a Somme-type assault, the Australians (and Americans) captured all their objectives, with the German defenders dead, wounded or captured. The skilful co-ordination of overwhelming firepower – heavy guns, field guns firing shrapnel, tanks, Lewis gun teams and grenade-throwing teams, rather than lines of massed Lee Enfield-equipped infantry, shoulder-to-shoulder – worked brilliantly. This successful Fourth Army concept was a beneficial culmination of the ideas and experience of junior and senior commanders combined, coming together with the ready availability of considerable numbers of guns and tanks, and plentiful supplies of ammunition.

The challenge for these tactics was to keep the flow of the offensive going, after the initial attacks had been launched. Artillery would need to register on new targets and do so day by day. Understanding how to maintain the momentum of the advance while still co-ordinating all aspects of combat power – infantry, artillery, tanks, combat engineers, logistics and aircraft – became the army's greatest challenge.

The success of the attack on Hamel allowed Rawlinson the opportunity to consider a large-scale offensive designed to protect Amiens, rather than merely expending his effort in building up his defences. During July 1918 a succession of raiding parties from the Australian Corps brought back incontrovertible evidence that the German defences in the region were lacklustre. Accordingly, he developed a plan to break through the Germans positions on 8 August. Using much of the warfighting methodology proven at Hamel, his plan was for ten divisions and 552 tanks to attack on a 12-mile front. In the first phase, four divisions of 26,000 men (one British north of the Somme; two Australian and one Canadian south of the river), would advance in their assault platoons supported by tanks (six per battalion) to seize the German front trenches and then to move beyond this by between 1,000 and 2,500 yards. The infantry would attack in columns, the shoulder-to-shoulder field tactics being consigned to history. The men would carry only what they needed in the assault: Lewis gun magazines and grenades. Enemy machine-gun positions (wired trenches or concrete emplacements) would be attacked from the rear or the flank by machine-gun teams and mortar teams operating in unison from different sides, while being engaged with fire from tanks. Pockets of resistance would not be attacked by infantry but left to be bombarded to obliteration (or submission) by mobile field artillery moving forward with the infantry.

Once these positions had been made secure by the first echelon of assault troops the second group of four fresh divisions would move beyond them – accompanied by the tanks, supported by 30 supply tanks, carrying extra fuel and ammunition (and, in some cases, carrying the infantry) – to capture the second objective, a further 3,000 yards beyond. In this phase the tanks would lead the advance, supported by a creeping barrage of high explosive, with the infantry mopping up behind. Once these positions were secure, this second echelon would continue the advance for a further 1,000 yards to secure the outer Amiens defence line. Simultaneously, a further two divisions would secure the high ground on the southern flank of the main attack. As at Hamel, the element of surprise remained the centrepiece of the plan. Preparations were made in conditions of unprecedented secrecy. Guns pre-registered on their targets by means of maps only and dummy radio stations were established to conceal the southward movement of the Canadian Corps.

For the first time in the three and a half years of war, all warfighting components were operating in unison within the context of a sensible plan, co-ordinated by intelligent, experienced commanders and executed by experienced, confident troops. Distinct, though complementary, roles were undertaken by field and heavy artillery, tanks, and machine-gun-dominated infantry assault platoons. The movement forward during the battle of combat supplies in supply tanks would enable the tanks and forward infantry to be replenished between the two assault phases to ensure that momentum was maintained. The campaign plan was based on the need for surprise and for the sequencing of the phases of battle. In addition, 365 aircraft were used to drive away enemy aircraft, provide observation of ground targets (such as artillery batteries), maintain communication (through message dropping to forward troops) and attack high-value targets such as enemy artillery batteries, troop concentrations, ammunition depots, railway marshalling yards and so forth.

So much had changed in the two years since the Somme. Aircraft observed the battlefield from above, spotting for artillery and observing troop movements, passing back information designed to influence the immediate battle. Engineers built minefields to channel enemy assaults into killing areas: in the offence, they cleared the enemy's mines and wire. Although tanks were still in their infancy, infantry were being brought forward under the protection of tanks, and sometimes even in them. There had been a revolution in artillery. In 1918 the British artillery was phenomenal in its scale and sophistication. Rather than, as in 1916, trying to obliterate enemy trenches prior to an infantry assault, field artillery was used to provide creeping barrages with shrapnel to protect advancing infantry, and heavy artillery fired high explosives against strong points, artillery batteries and critical defensive infrastructure. The army now knew exactly how much artillery, of each type, was required. In retrospect the total of 202 4.5-inch howitzers available to Rawlinson in 1916, the equivalent of one gun for every 100 yards, was woefully insufficient. However, for the Amiens offensive on 8 August 1918 Rawlinson had 700 guns and 350,000 rounds of ammunition. Counter-battery fire was now an art form, with sound ranging, aerial photography and flash spotting identifying the exact locations of nearly every enemy battery. More importantly, a conformity of training had been asserted across Fourth Army, with the result that high standards were expected, and attained, on every gun position.

In a way unforeseen even two years before, it was the sophisticated co-ordination of all these elements of combat power – infantry, artillery, air power, armour, engineers and so on, working closely together within a single operational (usually divisional or corps, but sometimes also army-level) plan – that was to deliver success on the battlefield in late 1918. The intelligent co-ordination of combat power – warfighting – meant that unreasonable pressure for battlefield success was lifted from the shoulders of the poor benighted infantry. But it meant too that armies were enabled once more not merely to move on the battlefield, but also to *manoeuvre* (i.e. to so deploy the integrated elements of combat power in time and space to achieve a successful campaign plan, such as the breakthrough of an enemy defensive position, like the Hindenburg Line, for instance). Modern warfighting skills, technology and methods secured victory for the British Army and its allies in 1918.

On 8 August Fourth Army began its offensive against the Hindenburg Line with 75,000 men against about 37,000 defenders. The German defensive preparations were poor. The Hindenburg Line was a linear defensive structure, not built to great depth, so negating all the advantages the Germans had demonstrated by their defences on the Somme. Artillery batteries were in fixed positions, which made them vulnerable to British targeting technologies. British logistical arrangements were astonishing and spoke to a war-winning level of administrative sophistication. A total of 290 trains brought up supplies and reinforcements. It was, as one participant, Captain Charles Carrington, described it, 'at last the battle of a dream, the combination that we had all longed for of modern techniques with fighting experience.'[2] Skilful British tactics helped diminish German morale and even fresh German divisions were unable to resist the British attack. Unlike 1916 and 1917, infantry soldiers in 1918 were able to overwhelm the targeted enemy trenches before the Germans could react. Fast-moving combat teams of mixed infantry and tanks would advance quickly through and between German positions. Enemy caught in woods would be surrounded and pummelled by artillery, without resorting to an infantry assault, which was the tactic of a previous, less enlightened era. Command of the air and the ability to pre-register the British guns on each of their targets meant that the Germans were never given warning of an attack. Beyond the initial targets and the end of the creeping barrages, the infantry assaults continued unhindered, the tanks dealing successfully with German machine-gun posts and infantry defences. Counter-battery operations crushed the German artillery at the outset of the battle. As a result, the British achieved all their objectives within 12 hours for the relatively low cost of 9,000 casualties. The Fourth Army advanced on a 15,000-yard front, captured over 450 guns (from 500) and inflicted 27,000 casualties, destroying six German divisions in the process.

Not everything went to plan on 8 August 1918. Where the attackers went beyond their field artillery support and the Germans were able rapidly to insert their reserves into weak spots in their defences, the attacks were held up. In the British III Corps area, north of the Somme, much went wrong. Its two assault divisions had been weakened and exhausted by two previous days of fighting, taking back a sector of the front in a counter-offensive. The ground across which they were to operate was water-sodden, the corps receiving only 36 of the 288 tanks available. Likewise, in this sector the counter-battery offensive was far less effective than that achieved in the Australian and Canadian sectors.

The second and subsequent days were difficult. The artillery and its ammunition needed to be brought up, and none of the new targets along the front had been pre-registered. The carefully co-ordinated relationships and synchronisation between infantry, artillery, combat engineers (for river and canal crossings) and tanks had been broken by battle, and now needed to be glued back together again. Communications networks were no longer certain the further the army advanced, as no line had been laid, so it was difficult for units and formations to stay in touch with each other. Divisions could not contact brigades, and brigades struggled to stay in touch with their battalions. The complicated logistics required to feed, rearm, reorder and repair front-line troops and key assets (such as tanks and guns) had to work itself through.

What was even harder was ensuring that units and formations could be adequately prepared to continue the offensive, with plans made and communicated to units well in advance of H-Hour. In the days that followed a certain amount of chaos was the order of the day as new lessons were learned while Fourth Army tried to keep the offensive going. Co-ordinating an entire army in the attack on the hoof when that army had not moved since it had been created in January 1916, was always going to be very difficult. Communications broke down; artillery couldn't get into its new positions in time; written battle plans and instructions didn't arrive on time. In the meantime, the Germans had managed to bring up their reserves, to plug obvious gaps in the line. Nevertheless, these reinforcements were themselves often rushed into battle piecemeal, without their accompanying artillery, and few were sufficiently prepared and trained to launch counter-offensives against the British. That is, even if they knew exactly where they were, because the front line had now changed, and in many places it wasn't certain where it was or should be.

The big challenge for the British now was whether Haig or Rawlinson would realise that all the elements of success on the first day – 8 August (creeping artillery barrages and targeted counter-battery work; tanks and infantry assault teams) – would need to be retained as the offensive focus for all subsequent days of an offensive. These techniques were, therefore, key elements of every attack, not just the *first day* of an offensive.

The next set-piece attack by Fourth Army was scheduled for 14 or 15 August, but was called off when Lieutenant General Arthur Currie, commanding the Canadian Corps, demonstrated that the strength of the German positions ahead of him – especially uncut wire – would make another offensive, without the benefit of the preparation they had managed to secure for the initial attack on 8 August, impossible. Rawlinson and Haig agreed, and the Battle of Amiens was at an end.

During the remainder of August, through to the Armistice on 11 November 1918, Fourth Army, together with the other armies of the BEF, made remarkable progress against even the most formidable German positions, demonstrating that the Allies had discovered – not without pain – the secret to successfully destroying enemy positions, overcoming even the strongest defences, and returning mobility to the battlefield. The progress of the offensive can be summarised as a set-piece battle, with the co-ordination of the artillery, tank and infantry as had been demonstrated to be successful during the Amiens breakout. The challenge that remained was to work out how to react quickly to changes on the battlefield and to reorganise at speed. What was required was for armoured, mobile field artillery to be able to accompany the tanks on to a new position and provide a creeping barrage for the infantry on to the next stage of their attack. Likewise, armoured support vehicles were needed to bring up combat supplies, and a communication system had to be developed that allowed commanders to remain in touch with their subordinates during the battle. This was essential to enable them to plan the next stage of the battle knowing where the enemy was relative to their own troops, and the strengths and weaknesses of both. But at this stage of the war, it was not possible to sequence the battle on the hoof. The only way it could be done was to pause, regroup, plan the next stage (which included registering the artillery on new-found gun positions) and issuing orders accordingly. It was *methodical* rather than dynamic, but it was successful for all that. The new approach to battle repeatedly overcame even the strongest German defences manned by the freshest troops. The secret remained the close co-ordination of all arms – infantry platoons reconfigured around machine guns, mortars and rifle grenades advancing under the cover of carefully choreographed creeping barrages provided by 18-pounder field guns; heavy guns smashing enemy batteries so as to prevent counter-fire against the British assault; and tanks supporting the infantry assaults by removing the threat of enemy machine guns. The advance was always co-ordinated with protection given to the flanks to counter the threat from enemy machine guns. Considerable logistical effort was made to bring up ammunition directly into the battle zone to replenish the field artillery.

But by far and away the most significant aspect in the new tactics was the accuracy and power of field and heavy artillery. In virtually every instance where the artillery was used in its revised role, attacks were successful, even when not accompanied by tanks. Pre-registered guns were able to provide an element of surprise. Even where preparatory bombardments were required, such as on the Hindenburg Line, the new form of obliteration was vastly different to 1916. Wire was cut and enemy positions destroyed by heavy bombardments using high explosive, not shrapnel, and to cut channels into the approaches that the attacking troops would take, rather than undertaking equal attacks across the entire front. Where infantry were required to advance without the support of artillery, whether by accident, oversight or misguided assumptions, they suffered badly from enemy machine guns and artillery. It is not the case that these victories were brought about solely because of increasing German weakness. In every position they could, the Germans fought with everything they had, but were undone by the sophistication of the offensive approach adopted by Rawlinson and his army. A significant aspect of the new type of warfighting was the British reluctance, by the start of what became known as the Hundred Days battle, for manpower to be sacrificed unnecessarily when firepower could be used instead. British commanders were now carefully husbanding their infantry, refusing to attack when supporting artillery was not available or adequately planned or prepared.

The British Army in 1918 won its battles not at the expense of the lives of its soldiers, but owing to tactical skill based on a mastery of an artillery-dominant offensive in which machine-gun-equipped infantry worked closely with tanks, combat engineers and aircraft, deploying a sophisticated approach to warfighting in which a single weapon was crafted from its many constituent parts. It was built on the bloody experience of battle. Though mistakes continued to be made, most British commanders demonstrated that they were able to understand and master the intricacies of the modern battlefield. The industrialisation of fighting had rewarded the British, whose soldiers by 1918 had all the equipment they needed. Likewise, British, Commonwealth and American soldiers never lost their willingness to continue to fight, unlike the French who suffered a number of mutinies in 1917 and struggled to rebuild morale.

By the end of 1918 the British Army had mastered what might be described as 'methodical' warfare. It understood every constituent element required to undertake successful battle in a step-by-step manner. Methodical battle was designed to unpick the enemy's defensive lock one piece at a time. The fast mechanised (and sometimes armoured) troop transport, and mobile artillery, supported by ground support aircraft such as tactical bombers and dive-bombers, all glued together by radio communications, were the next advance in tactical and operational thinking. This would, in a future war, enable more dynamic or fluid operations, where decisions could be made spontaneously as battle took place. It would allow for smarter warfighting. Dynamic warfare, by contrast, allowed for the attacker to look at other ways to defeating the enemy operationally, rather than attacking their defensive positions head on. Dynamic warfare allowed the attacker to identify positions, infrastructure or localities that were critical to the enemy's ability to continue fighting. Strong defensive positions could therefore be bypassed and allowed to wither on the vine, if a more strategic target, such as a headquarters location, railway junction or supply depot behind the front lines, offered itself. It might be that battlefield superiority could be achieved by bypassing the fixed defences entirely, if this would demoralise the enemy and force them to turn back or attempt to reorganise themselves to defeat a line of attack they were not expecting.

This discombobulation was key to German success in northern France in 1940. The Great War did not allow for dynamic warfare, but it did allow for the expert development of methodical warfare, its essential precursor. Such warfare was entirely mastered by British and Commonwealth armies in 1918. This learning was captured in a series of excellent doctrinal pamphlets after the experience of July 1916, revised regularly, that ensured that a vast citizen-based army was 'singing from the same hymn-sheet'. The seminal *Instructions for the Training of Divisions for Offensive Action* was published in December 1916, and laid out in great detail how formations needed to train for the type of connected, integrated battle experienced on the Western Front, where close and careful co-ordination between infantry, artillery and tanks was essential to success. Doctrine is required from bottom up as well as top down and in this respect the most important document to be published was *Instructions for the Training of Platoons for Offensive Action*, which appeared in 1917. It saw several iterations through to the end of the war and would have been the document which guided the training and deployment into battle of Second Lieutenant Duff Cooper, whose experiences we have already recorded. At divisional level the doctrinal handbook was *The Division in the Attack – 1918*, published just as the war came to an end. It was a document of profound good sense, reflecting just how successfully this citizen army had developed its theory and practice of war in the two years since the Somme. The battle that Cooper experienced was perfectly encapsulated in the description it provides about the philosophy of decentralised command:

> The successful conduct of a battle depends upon the rapidity with which local successes are gained and exploited. As the advance proceeds and the enemy's organised defences are overcome, the actual direction, and to a large extent the control, of operations must necessarily devolve upon the commanders on the spot. It is absolutely essential, therefore, that commanders of all grades should be able quickly to grasp the salient features of a tactical situation and to act with boldness and decision.[3]

The Hundred Days taught the British Army that at the heart of *warfighting* lies the premise that successful campaigns strike not just at the enemy's military mass and its physical capability (its tanks, infantry, artillery, air power and so on) but also its ability to command and direct its forces, as well as the morale of both commanders and men, and their individual and collective will to fight and win. It reflects the need of an army to adapt to meet the ever-changing demands of a complex battlefield. It also reflects the need for a command philosophy that allows subordinate commanders at all levels of the rank structure to understand their commander's *intent*, while executing operations according to the local circumstances, but always within a set of sensible rules laid down in the fighting manuals. Herein lay battlefield success. The successive victories in 1918 didn't just unravel Germany's military power, but attacked its self-belief, persuading many that their sacrifice would be wasted were they to fight to the death. An army does not need to be defeated in detail on the battlefield, but its soldiers need to believe that they have lost. That army is indeed then beaten.

It would take the next war for dynamic warfare to be fully developed. It would be mastered in the first place by the losers in 1918 – the German Army. The British Army too quickly forgot what it had so painfully learned.

It is ironic, therefore, that from 16 July 1918 the primary staff officer in one of Rawlinson's divisions, 47th (London) Division on Fourth Army's northern flank, was the 30-year-old Lieutenant Colonel Bernard Montgomery. As the battles developed during 1918 Montgomery, responsible for deciding upon and designing the plans for battle, became a convert to the tightly organised methodological approach to warfare that Fourth Army had developed and would use to great effect in the battles that followed, and which finally broke the German Army in Flanders. Yet it was not until the Second Battle of El Alamein in late 1942 that 'Monty' was to demonstrate that he had inherited the legacy of 1918.

But to return to mid-1918, the army in which Duff Cooper and the men of 10th Platoon were a part was at the top of its game. Victory for the British Army in 1918 was brought about as a result of the tactical skill of its well-equipped and well-trained machine-gun-equipped infantry working closely with artillery, tanks, combat engineers and aircraft, deploying a relatively sophisticated approach to warfighting. It was built on the bloody experience of battle, as this warfighting methodology had been developed directly as a result of what had been learned day by day on this new type of battlefield. Although mistakes continued to be made and men continued to die, they did so in relatively fewer numbers and for greater purpose. Most British commanders demonstrated that they were able to understand and master the intricacies of the modern battlefield. The British Army was, in late 1918, an impressive, war-winning weapon.

The cost was high – 512,600 British dead on the Western Front alone at the war's end – but at the time this was widely considered an acceptable exchange for victory.* It was, as Corelli Barnett reminds us, 'much lower proportionate to population than the German loss, and not much more than half France's losses'.[4] One might wish the process of learning to have been more rapid than it was, but the accusation that the generals were dunderheads who had no answer to the slaughter of the Western Front is undone by the fact that, although many dunderheads fell by the wayside, it was the generals who were in command in 1916 who worked to find a way out of an impasse they and their predecessors had not seen. That the process was to take the lives of so many millions was part of the tragedy of industrialised warfare.

*The total British Empire war dead approached one million.

PART TWO

Post war and inter war

Chapter 5

Peace, and derangement

The end, when it came, was with a whimper. At 6.50am on 11 November 1918 the duty staff officer at the British Expeditionary Force General Headquarters at Montreuil-sur-Mer, Lieutenant Colonel William Dobbie, received news that an armistice had been signed an hour before between the Allies and the Central Powers. He immediately dictated a signal to all troops under Haig's extensive command:

> Hostilities will cease at 1100 today, November 11. Troops will stand fast on the line reached at that hour, which will be reported by wire to Advanced G.H.Q. Defensive precautions will be maintained. There will be no intercourse of any description with the enemy until the receipt of instructions from G.H.Q. Further instructions follow.

It had been clear to many observers for several weeks that the German Army was close to throwing in the towel. Captain Charles Carrington was certain as early as Ludendorff's offensive in March 1918 that the Allies would finally win. He and his friends had never been in the 'slightest doubt about our war aims, which were to drive the German Armies out of France and Belgium and to give them such a pasting in the process that they would not again be able to make unprovoked assaults on us or our neighbours.'[1] This is what the BEF was now doing. They were winning. The pacifism through which society was later to view the war came, he believed, from civilians who saw only the horror of death rather than the triumph of victory. The war came to be seen afterwards as a natural disaster which bred a new scepticism for the old certainties. It was a pacifism directed at the army – rather than the Royal Navy – as it was the army and its generals who had so urged the pre-war commitment to an alliance with France, and were seen as uniquely guilty of what happened next. That summer Carrington, a hardened veteran of the trenches, read one of the first books of the disillusionment school (*Le Feu*, by Henri Barbusse) and suffered nightmares as a result. Imagination, it seemed, was more terrible than reality. By September, with the daily news of Fourth Army's victories, it was clear that the war was all but over. There was no way the Germans could survive this systematic destruction of their fighting power.

On 13 October Carrington wrote to his parents in far distant New Zealand about the rumours that were emerging almost every day that the Boche had surrendered. On the night of 7 November, the German emissaries received the terms of surrender. They signed, in the railway carriage at Compiègne, soon after 5am on 11 November. The moment the troops received the signal, at breakfast time, the whole Western world, so it seemed to Carrington, dissolved into a carnival that did not slacken for 48 hours.

The peace in 1918 was nothing like the victory that was attained in 1945, in the way it was perceived in the public mind and how it was appreciated in terms of the public mood. Certainly, the war had stopped. But there had been no build-up of expectation of war's end, no public appreciation of the weakening of the German armies and of the long-term, strategic failure of Ludendorff's March offensive. When the end came, it was an anti-climax.

Compared to 1945 the immediate celebrations were initially muted. A crowd gathered at the front of Buckingham Palace on the afternoon on 11 November, waving flags. That night a display of captured German guns on Horse Guards Parade was dragged by a crowd to Trafalgar Square where they and their wooden carriages were burnt beneath the plinth to that previous vanquisher of continental militarism, Horatio Nelson. Elsewhere across towns and villages, church bells clanged their relief. But there was no organised bunting or parades, victory being so unexpected. The parties began the following day, the relief palpable in the exuberance of celebrations which Charles Carrington likened to a 'succession of bank holidays or boat-race nights'.

In France, Carrington had been given orders, two or three days before, to proceed to the railway at Le Havre station. He was to leave with a draft of 1,100 soldiers for Italy at 11am on 11 November 1918. Not a superstitious man, he found the coincidence amusing more than anything else. War or no war, orders were orders. In fact, the certainty of purpose the orders provided was helpful in the confusion of the moment. What would happen next? For the meantime, the battalion was off to Italy, and it didn't matter:

> Our 1,100 set off in a state of mild hysteria, to march from Harfleur Camp to the goods entrance at Le Havre railway station, arriving at midday among crowds gathering in the streets, who sang and waved flags, or offered our men drinks from sidewalk cafes, while girls broke into the ranks and marched with them, arms enlaced.

The officers spent the day dragging the men from the *estaminets* to ensure that the draft left on time at 4pm. The journey, at a leisurely ten or twelve miles per hour, took them through sunny weather along the Marne. The soldiers, accommodated in straw-strewn horse carriages, now with no enemy to fire at, began a little target shooting of their own to celebrate the peace. After a while they grew bored of even this and, leaving a trail of dead cows and injured farm buildings behind them, resorted to absconding as the train slowed to pass through stations along the Rhone, to enjoy the attentions of the cheering crowds on the station platforms. They would then catch up with the troop train by hopping on the 'Mail', which travelled at four times the pace.

With the war now behind them, Carrington's battalion settled down to an unexpected peacetime routine in Italy. Life was pleasant. The regimental band played every day for the guard mount, watched by friendly locals. The soldiers had a well-stocked canteen, the officers a fine mess. Films were shown nightly, the divisional concert-party put on weekly pantomime performances and education classes were laid on for soldiers in preparation for the fast-looming reality of civilian life. They played sport every afternoon and enjoyed organised trips into Venice. All of a sudden, the British Army had become tourists abroad, awaiting their transition as individuals into civilian life. Only military discipline and the comradeship of men who had been through the fire of the trenches held the men together. Discipline was therefore applied with a light touch: troops drilled and trained in gleaming equipment every morning and had the afternoons off for sport and leisure.

The fast-encroaching realities of civilian life intruded into their lives that December with a general election in which David Lloyd George was re-elected with an overwhelming majority for his coalition (Conservative and Coalition Liberal) government. This election reflected that the political geography of Britain was changing. For the first time all adult males, together with many adult females, were able to vote, increasing the number of voters from seven to 20 million. The franchise had thus expanded considerably, and few were interested in exchanging the financial benefits of peace with the endless cost of war. This was clearly a good thing in the development of Britain's democracy, but with so many other pressing priorities these new votes weren't necessarily ones for defence. Politics was moving quickly from the elite to the popular, with parties and politicians now concerned more than ever before with public opinion.

For those who had had a job before the war, the process of resetting one's expectations to 'civvie street' was not difficult. The problem came for those who had had no such experience. Carrington was scared of the prospect. 'What was "de-mobilization", the new vogue-word, to the likes of me?' he asked himself. Soldiers were allocated a work category with a priority number supposedly representing the needs of the economy. Carrington found himself at the bottom of the pile, with a code-number of Class 38, 'student'. He wasn't in a hurry to go home, but others were. The problem was fairness, an issue which the British soldier took very seriously. Why should those who had served longest at the front be beaten home by Johnny-come-latelies who had barely had time for Flanders' mud to stick to their boots? Discontent became pronounced. In battalions and regiments, it was kept under control: in depots and base camps where the soldiers didn't know each other, let alone the officers responsible for what had been until recently the largely administrative business of getting men to and from the front, it could turn ugly.

Although the principles of a demobilisation plan had been agreed in Whitehall in 1917, the practical arrangements were not well thought through. The dismembering of the army that had brought victory was haphazard and did not augur well for the future health of the army. Haig had suggested an alternative plan to the one that was based on releasing the men according to what the economy required. He needed to retain a substantial part of his army for the purposes of occupying a defeated Germany. For this he would need 16 of his current 60 divisions. He wanted the priority for the drawdown of those troops superfluous to requirements to be based on time in service, military record and war injuries. It was a soldier's rather than an industrialist's plan, but it was rejected. Whitehall's decision ignored – to its cost – the reality that at the end of the war large numbers of men would want to go home, quickly, and that soldiers would not accept something that was unfair to them, the servicemen who had carried the heat of the day. The bureaucrat's plan was patently unfair to those who had served or suffered the longest. Understandably, the approach did not sit well with men who had been fighting for several years, and who considered their own needs for repatriation to be as important as industry's. First in, last out simply wasn't fair, even if the bureaucrats in the Ministry of Reconstruction thought it efficient.

A significant crisis of discipline affected parts for the army, with a mutiny of sorts starting in Folkestone on 3 January 1919 and others following in London and Glasgow. Soldiers in uniform marched somewhat chaotically in London with placards that said 'We won the war. Give us our tickets.' A more serious disturbance occurred in a rest camp in Calais on 27 January 1919 by three or four thousand men whom Carrington disparagingly referred to as 'base wallahs'. Seizing weapons, they defied the authorities. London, all too aware that the revolution in Russia had been led by angry soldiers, realised that urgent action was required. Haig immediately sent in two fighting divisions and threatened to open fire if the rebels didn't surrender. The mutiny collapsed. But Haig had been right: large-scale disobedience could follow inept handling of fighting men by civilians and bureaucrats. Large numbers of well-trained and fully armed disaffected soldiers potentially constituted a fundamental threat to the state. If Britain was to avoid the mutinies which had ravaged the French Army in 1917 a more enlightened approach was required. Churchill, newly appointed as Secretary of State for War, in conjunction with Haig, quickly reversed Whitehall's decision. From henceforth, demobilisation would be conducted on the basis of length of service, age, marital status and wounds.

In the 'Explanatory Note by the Secretary of State for War' on 29 January 1919 Churchill observed that of the 3.5 million in the army on 11 November, 750,000 had already been discharged or demobilised under the old regime of 'industrial priority'. In only two and a half months a fifth of the army had already been demobilised. He anticipated that an army of 900,000 men would be required for the immediate post-war tasks associated with ensuring the peace. The remainder would be released just 'as fast as the trains and ships can carry them and the Pay Offices settle their accounts'. The rate of pay for serving soldiers was raised and bounties were offered to men who agreed to enlist as, even with the dramatic reductions called for by the ending of the war, a churn of new recruits would be required to sustain the new peacetime army. The tasks required of these 900,000 covered five specific areas: the Home Army; the Army of the Rhine; the Army of the Middle East; the Detachment of the Far North (i.e. a commitment to Russia, described later); and garrisons of the Crown Colonies and India.

Churchill's statement concluded that in 1919 'we must remake the Old British Regular Army so as to provide on a voluntary basis the Overseas Garrisons of India, Egypt, the Mediterranean Fortresses and other foreign stations.' The intention was that 'we shall not require to keep so large an Army as 900,000 in the field', although, quite sensibly, no estimate of the ultimate requirement was conjected. Time was shortly to show that this was nearly three times higher than the number the government believed it could afford and settled on in 1923. Although Churchill's 'Note' related to the drawdown of troops resulting from the end of hostilities in Europe, it is instructive in terms of the overall expectation of the need for troops for warfighting purposes on the continent of Europe after the war. The commitment to such a force is absent, though it could reasonably be assumed to be included in the troops designated for either Home Army or Army of the Rhine duties. At the same time, across a chaotic and unstable Europe everyone turned to the Allies to help with 'pacification' duties. Although the British Army was in great demand, the country 'could not spare many troops and could not risk sending small detachments to remote districts far from the sea. All these consequential war measures occupied during the first months [of 1919] much of the time and the energy of the principal Powers.'[2]

Within a year the armed forces had released about 3 million men from khaki into civilian clothes. In 1918 there had been 4,583,300 servicemen and women in the British Armed Forces, representing 10,521 personnel for every 100,000 of the British population, the highest in the country's history, exceeding even that at the height of the Second World War when it was 10,006 per 100,000. By the end of 1919 this figure had fallen to 1,604,700, a drop of 65 per cent, now representing 1,388 per 100,000. The following year it fell again to 597,700 and in 1923 it reached its nadir at 334,700, a 93 per cent drop on manpower over four years, representing 753 servicemen for every 100,000 of the population.[*] It did not fall to

---

[*]The regular British Army had stood at 250,000 in 1914.

this number of service personnel again until 1977, by which time the withdrawal from empire was almost complete. For all the right reasons, demobilisation savaged the British Army, ending up with a manpower complement far below the 900,000 that Churchill had initially assumed would be required. As part of this dramatic reduction in manpower, changes took place to force structures as well. The Machine Gun Corps was disbanded, its Vickers machine guns being parcelled out in groups of eight to create new machine-gun platoons in the remaining infantry battalions. The newly raised Tank Corps, which had numbered 25 battalions at war's end, had been reduced to five by 1921. Indeed, it was lucky to survive on its own at all.

1919 failed to turn into a year of jollity now that the war had been vanquished, the swords beaten into ploughshares by the peace. For many the dampener on the public mood brought about by the war was caused by the relentless and gargantuan demands of public and private sacrifice reflected in the endless casualty lists published daily in the newspapers until 1917 (thereafter weekly by His Majesty's Stationery Office). In Churchill's account of the war, first published in 1929, he used interesting language to describe how the sudden, unexpected end of the war affected Britain. Planning for the future of the army, especially not for a resumption of the sort of hostilities that had recently ended, was not on anyone's agenda. 'Not only the armies but the peoples [of Britain] were profoundly affected by the sudden cessation of the war,' he observed. 'The poise and balance even of Britain was deranged.'[3] Derangement indeed. The winter of 1918 was cold, and coal was short. Two waves of the 'Spanish influenza' that was to infect a quarter of the world's population swept across the country, causing the deaths of 228,000 Britons. A third wave struck in the early spring of 1919. At the same time industrial unrest shook the country, with 2.4 million workers going on strike during the year, with especial militancy in the mines and on the railways. It was a particular problem in Scotland where the

Battle of George Square was played out on 31 January 1919. The red flag of bolshevism on the municipal flag pole was enough to worry the authorities that revolution was imminent. But Glasgow was not Petrograd, merely the locus for the re-inflammation of fragile British industrial relations following the end of the war. While troops were deployed in 'aid to the civil power' it was police baton charges that felled the workers. It led to the sheriff reading the Riot Act, after which the army marched, the city centre being occupied by troops in helmets and fixed bayonets. That particular battle was quickly over but strikes and threats of strikes became a recurring feature of the country through to 1921.

Fortunately for the government, British trades unionists were not seeking the overthrow of the country, merely asking for better working conditions and salaries for their workers. There was an enormous difference between Petrograd in 1917 and Glasgow in 1919, but the clanging cymbals of civic unrest were unsettling and discordant. The fact that the army was marching through a British city, supported by tanks, only 82 days after the end of the war, made those of a nervous disposition fear that the spectre of war had diverted its gaze homewards, rather than in other parts of the world where wars normally took place. Anxiety about communism gripped many in government, fearful of a threat to the survival of a liberal political settlement that had been centuries in the making and offered a re-run of the terror in France 130 years before.

The mood lightened when the Treaty of Versailles was signed. Victory Day was announced for Saturday 19 July 1919 (so as not to disturb the working week) and the British Army was able to parade, colours flying, through the capital city of the empire. Charles Carrington watched the show:

> London was decorated and thronged with visitors as for a Coronation. A great camp was formed in Hyde Park for representatives of every regiment in the British Army, every Dominion, every Ally. The procession of troops bearing all their regimental colours marched nine miles through the East and West Ends of London to salute the King at Buckingham Palace. They

passed through crowded streets which cheered for Foch, who flourished his British field marshal's baton, and louder – as I heard with my own ears – for Haig. No doubt we had won a victory.

Despite the flags and parades Britain's moral certainties had been holed below the waterline. Not yet visible perhaps, but in a few years, these would exhibit themselves in widespread revulsion at the possibility of another continental war. Thinking of, or preparing for such, was social and political suicide. The idea of another war was so anathematic that arguments for preventing it by means of deterrence or of defensive plans or preparations were considered morally suspect and political suicide. Support for pacifism was to reach unprecedented levels in the decade that followed. The catchphrase of the moment was 'never again'. Even Sir George Milne, Chief of the Imperial General Staff (CIGS) at the time, in 1926 referred to the recent war as 'abnormal'. He did not consider it likely that Britain would ever be required again to fight a European war. Those few who considered such issues were quickly dismissed as militarists or worse, warmongers. Lieutenant General Sir Francis Tuker was later to describe this widespread attitude as 'mental disarmament'. This involved 'the nation frowning upon any who thought the use of force might in some circumstances be right,' he argued. There was:

> a definite and abusive discouragement of those who dared at any time to write about a possible war and about the nature it might assume. There was, therefore, no study of war in the nation itself … If as a soldier he did study it, then he was at once labelled as a Blimp and was regarded by the civil population as something of a criminal.[4]

Rudyard Kipling was on to something where, in the final verse of 'Tommy', we hear:

> For it's 'Tommy this, an' Tommy that, an' Chuck him out, the brute!'
> But it's 'Saviour of 'is country' when the guns begin to shoot …

The thought that the 'guns begin to shoot' again was an alien thought. The world in 1919 was one both of relief – at being alive – and disillusionment with the realisation that the modern, civilised world could have been so foolish as to venture into such a destructive war, in which so much had been lost. Surely, this could never happen again in Europe? But it did in 1939 and again in 1991 in Yugoslavia, 1999 in Kosovo and 2022 in Ukraine.

———

Indeed, the end of the war against the Central Powers did not see the end of all wars, as many hoped it would. Abroad, the twin revolutions in Russia in 1917 and the ongoing civil war were reminders of the fragility of the peace that had settled suddenly over Western Europe: the East was still in turmoil. The principal British involvement in Russia* came in July 1918 in an attempt by the Western Allies to ensure the protection of a Czechoslovak army operating along the Trans-Siberian railway. They were fighting the Bolsheviks who had launched a revolution within a revolution the previous October. An Allied force of US, British, Italian, French and Japanese troops – under Japanese command – entered Russia from Vladivostok and advanced to establish itself in the region of Omsk. Meanwhile, a further 7,000 mainly British soldiers disembarked at Murmansk and Archangel in March 1918 initially to prevent the vast quantities of supplies the Allies had provided to the Russian Army from falling into German – or Bolshevik – hands. The intervention was not so much to defeat the Bolsheviks (despite what Bolshevik propaganda suggested) as to ensure that the Germans were not overly advantaged by the demise of fighting in the East. Although the purpose of the British and Allied missions throughout the years to 1920 was often opaque, as the situation constantly changed on the

---

*Three British armoured car squadrons had in fact been with the tsarist forces prior to the revolution, and remained there until late 1917.

ground, the commitment was significant: some 59,000 men at its height. The problem for the army was that it was a military commitment without a strategic purpose and, on that basis, an abuse of soldiers' lives.

During all of this turmoil and continuing commitments at home and abroad the army was still rapidly demobilising. The rapidity of this change, organised well after the earlier mistakes had been rectified, had a profound impact on society. During the 12 months following the end of the war nearly 3 million men flooded out of khaki, ten thousand each day in the first six months. Society, politics, the economy and the temperature of the nation were uncertain. What would the future bring? Would Britain be a home fit for heroes? Given what we now know of Post-traumatic stress disorder, the reality was that many of those coming back into civilian life following the sudden end of fighting would have been psychiatric casualties of one kind or another. A study published in the *British Journal of Psychiatry* in 2018 reported that 17 per cent of *combat* (as opposed to service) veterans from the wars in Iraq or Afghanistan reported experiencing PTSD.[5] In addition, 9 per cent of other servicemen and women who served in these war zones but not in a combat role suffered PTSD. The rate for the veteran population as a whole at the time was 7.4 per cent. It would seem fair to assume that with the relentless intensity of the combat during the Great War the PTSD rate was even higher than that reported in 2018: perhaps as much as 25 per cent. If this were the case, it meant that as many as 750,000 men returned home with levels of trauma that today would require some form of clinical intervention. Except for those formally diagnosed specifically with 'shell shock' (some 80,000) – surely a serious underestimate – no treatment was available for what was at the time an undiagnosed medical problem. War trauma, of course, affected not just servicemen. With over 700,000 British dead (9 per cent of the adult population under the age of 45) together with 1.5 million wounded, whole families – perhaps with extended families ten or twenty times this number – were left devastated by loss. The UK population in 1918 was 34 million.

It doesn't take complex arithmetic to recognise that in 1919 the whole of Britain could be considered as a psychiatric casualty of the war.

———

The Treaty of Versailles was considered by most people in Britain to end the possibility of further or future war in Europe. Just as the Concert of Vienna after Waterloo was followed by a century of peace for Britain in Europe, it was widely considered that a repression by the victors of the German militarism that had caused the war, and the division of the old Austro-Hungarian Empire into new nations reflecting their ethnic diversity and nationalistic aspirations would also lead (to quote a later statement earlier than it was made) to 'peace in our time'. This was what Britain wanted. Unfortunately, it was not what it received. It was assumed that Versailles would provide the peace while the League of Nations, a new and untested form of collective security, would preserve it. If both did their job, there would be no more war, as Lloyd George hoped, for at least 30 years. In these circumstances, the idea of military intellectuals planning on how to fight another peer-to-peer conflict on the scale of 1918 seemed absurd.

Charles Carrington recalled how he and other veterans of the Great War worried soon after the Armistice in November 1918 that no one at home among either the political class or the civilian population seemed to realise that the German Army had actually been defeated in the field. 1918 had seen the British Army's greatest victory. Indeed, as the historians Robin Prior and Trevor Wilson observe, it seems to have passed unnoticed by history that the spectacular series of victories won by the Fourth Army were not having to be paid for by 'an insupportable levy in the lives of their infantry'.[6] Even soldiers didn't seem to appreciate the point that they had won a tremendous victory, using new approaches to warfighting and against an enemy that had no effective *military* counter. The Germans hadn't lost just because they had been exhausted, but because militarily they had been out-fought in battle by a vastly

superior Allied army in which Britain had played a leading part.
Even then, in 1918, the seeds were being planted of the fallacious
argument that the German armies in France and Belgium had been
stabbed in the back. While Britain lulled itself into the sleep of
passivity, forces were already underway in Germany to believe this
nonsense, and to recover their lost glories. Carrington observed:

> The war was something rejected, forgotten 'by the civilians who
> saw it only as a disaster from which a new world was painfully
> emerging; and by the soldiers who saw it as an achievement,
> finished and tied away. The civilians wanted to hear no more of
> it; the soldiers kept it to themselves to be discussed in private,
> like a masonic secret.'[7]

In a strange, emotional metamorphosis the British public appeared
to wish to embrace and indulge the bloody memories of the Somme
and Passchendaele rather than the victories of 1918.

Chapter 6

Old and new post-bellum responsibilities – and the Irish Question

The Great War did not make the world a safer, more peaceful place. Nor too did its end in November 1918 result in any let-up for the British Army. A range of commitments, some old, some new and some entirely unexpected, impacted an army rapidly drawing down after the Armistice in accordance with Churchill's revised demobilisation programme. Indeed, the direct consequence of the Treaty of Versailles in June 1919 was to expand the British Army's commitments by transferring to Britain a number of 'mandates' from the League of Nations in respect of previous Ottoman and German-held countries and colonies. These were not easy to manage, as the revolt in Egypt in 1919 and the four-month revolt in Iraq in 1920 were to demonstrate, consuming vast numbers of troops just at the point when fewer were available because of the post-Armistice drawdown. Two additional divisions were hastily despatched to Iraq in July 1920 at precisely the time that greater demands were being made on the Army in Ireland. Then, in September, the CIGS, Field Marshal Sir Henry Wilson (he had been promoted in July 1919), warned General Nevil Macready, the Commander-in-Chief Ireland, that he needed ten battalions from Ireland (from the 51 stationed there) to help manage an expected coal strike in England. Macready warned that this would be catastrophic for the security situation in the country. Indeed, the period through to 1923 was one in which the respective Chiefs of the Imperial General Staff (Wilson until February 1922 followed by the 10th Earl of Cavan until 1926), struggled to make ends meet, endlessly recycling troops from various commitments around the globe to satisfy the most pressing demands as they arose. The military commitment to Russia continued beyond 1918. The revolution transmogrified into the Russian Civil War, British troops remaining in the rapidly disintegrating country because of fast-growing fears of a contagion of a Red revolution at home. Indeed, fear of the threat of Bolshevik revolution did not abate in Westminster until at least 1922, possibly not even until the end of the General Strike of 1926.

Industrial unrest was widespread. In April 1921 Wilson was busy planning for the prospect of a 'triple alliance' of strikes at the same time as abroad a range of commitments placed the army under pressure. Wilson even complained to King George V that he hated the 'dispersal of our little Army in Silesia, Constantinople, Persia, Mesopotamia etc when every available man was wanted in Ireland.'[1] Likewise, the Third Afghan War in 1919 was a sudden reminder that the Great War had not eradicated the pre-war security challenges across the empire. The simmering discontent in Ireland burst again into the flame of insurgency in the same year and if that weren't enough Mohammed Abdulla Hassan – the 'Mad Mullah' – in Somaliland initiated another revolt in his string of rebellions against the colonial power. Similarly, managing the aftermath of the Ottoman Empire also entailed the occupation of Constantinople and parts of western Turkey until a peace deal was secured with the Turks in 1923.

Each of these military commitments could be described as largely 'garrison' tasks where the presence of the army was to show the flag and bring stability through its mere presence on the ground. Where armed force was deployed it was concerned mainly with underpinning security for the maintenance of civil governance. In some cases, it entailed counter-insurgency operations. None of it was high-intensity *warfighting*, merely the often-humdrum fighting (or 'small wars') that the army has frequently been called upon to undertake in times of peace. But some of it – such as countering the insurgency in Ireland – required the army to undertake a radical shift away from what had occupied it during the recent war. It needed new thinking, new tactics and approaches to the use of military force, together with radically new training for troops. Counter-insurgency at home also represented a mental challenge for soldiers entirely different to that of putting down rebels in some dusty corner of the empire. In Ireland, the fight was closer to home. When it was over Major (no longer temporary Lieutenant Colonel) Bernard Montgomery, brigade major of 17th Infantry Brigade in Cork in 1921, observed that counter-insurgency 'is thoroughly bad for officers and men; it tends to lower their standards of decency and chivalry, and I was glad when it was over.'[2] The immediate resumption of a whole series of low-level though busy fighting tasks after November 1918 served to remind soldiers just how broad their job description was. Unfortunately, both political and military minds alike, overwhelmed with the sudden and pressing needs of the moment, forgot that the 'small wars' of empire were not the only ones the nation might be required to undertake.

No one in government showed any inclination, understandably perhaps at the time, for thinking about or preparing for another war of the kind that had just passed into history. Lloyd George's argument was that the country now had the right to spend the peace dividend its wartime investment had made on social betterment. In retrospect a finer balance was required than that which was achieved. Retaining an army large enough and capable enough to fight a war similar (but smaller perhaps) to that of the Hundred Days was not a vote winner even if, in hindsight, it was the right thing to do. In a direct challenge to the idea of retaining strong armed forces as an insurance to a future continental commitment, the prime minister observed that, 'as external threats to Britain had been eliminated if the country persisted in maintaining a larger Army and Navy and Air Force than we had before we entered the War, people would say, either that the War had been a failure, or that we are making provision to fight an imaginary foe.' Now that peace had been achieved, the task was to spend more on 'the health and labour of the people'.[3]

The counter-argument was made, but its impact was muted. A 1919 War Office-funded document entitled *The Mission of Britain's Army*, written by Lieutenant Colonel Hermann de Watteville, a gunner who had been an instructor at the pre-war Staff College, argued that a strong army did not mean that Britain was militaristic, but that it valued peace. In order to ensure peace, the country needed to be prepared for war. Pamphlet No 37 was a carefully argued defence of the famous dictum *si vis pacem, para bellum* ('If you wish for peace, prepare for war'). De Watteville's argument was that a strong army could serve to prevent war. It could even have prevented the outbreak of the Great War. Now was not the time to demilitarise completely. If Britain had had an army of 500,000 men, rather than the 80,000 who landed in France with the BEF in 1914, argued de Watteville 'the war might never have taken place.' He acknowledged that before 1914 'the inclinations and the views which were held by the majority of the British people did not tend that way' and an opportunity to prevent the war was missed. The country paid the price for this 'impolicy' in the blood and treasure lost during the war. Would it learn? Not if it dreamt that the war 'was a war to end war' argued de Watteville, or if it wanted a peace dividend or suddenly began to believe in the idea of universal disarmament.[4]

De Watteville's voice was one crying in the wilderness. There was simply no way that Britain, which had always eschewed the existence of a large, home-based standing army, would countenance one now, especially as the war was over. Even the War Office which had commissioned his pamphlet – paying him the princely sum of 20 guineas (about £1,200 in 2023 prices) – wasn't interested in anything other than resetting the clock to 1914. Indeed, the CIGS, Sir Henry Wilson, could not get anyone in government to think about what he called the 'Post Bellum Army' at all. The War Cabinet meeting on 13 November 1918 was 'the worst he'd ever attended', he noted in his diary. No one seemed interested in the future, instead enjoying wallowing in the euphoria of the moment.

The problem was that the country wanted to reduce expenditure on the army while at the same time dramatically increasing its commitments abroad. A month later Wilson produced a paper for the War Cabinet predicting a requirement for an army of up to 500,000 men to oversee the new commitments (and post-Versailles mandates) that the new government of Lloyd George required of the British Army. When Lloyd George at the War Cabinet meeting on 5 August 1919 asked why it was that the number of men in the army was planned to be 320,000 by spring 1920 when the pre-war number was about 255,000, he had to be reminded by Churchill of the vast new commitments his government had accepted following the end of the war. This was 'war imperialism', the accruing of territory for no other strategic reason than that it was gifted as the result of war, not even necessarily a war of one's own choosing.

To Wilson, who had to garrison (and fund) these various accumulations after Versailles, the lack of strategic sense in their retention was intensely frustrating. To his mind, there needed to be a rationalisation of overseas commitments, reflecting what was strategically vital to Britain. In the East this obviously included military bases in India: in the West, England and Ireland, followed by the 'Clapham Junction' of the empire, Egypt and the Red Sea. Everything else – including Iraq, Palestine, Persia and the Caucasus – needed to be accorded a strategic priority in respect of what it added (such as oil) to the security of the empire. If it added nothing, it should be deleted. Palestine, for example, 'didn't belong to us' and should never be Britain's responsibility; it was certainly not, to coin someone else's phrase, worth the bones of a single British grenadier. It would simply suck up troops, cost money, cause casualties and make people with a political axe to grind hate Britain for a reason not of Britain's making. Taking on the Palestine mandate was an act of self-harm, Wilson thought. He was fearful of being lumbered with providing a military solution to an intractable political problem.

Wilson's pragmatism did not succeed in winning over Lloyd George's emotional obligation, however, to a continuing military commitment to the occupation of Constantinople. The Prime Minister's Hellenism prevented him from doing anything other than supporting the creation of a strong Greece to support British interests in the north-eastern Mediterranean even though this meant a war with Turkey in September 1922. Egypt did in fact offer an alternative strategy. When independence was ceded to Egypt in 1922 the treaty recognised Britain's longstanding 'special interests' in the country, not least the Suez Canal, which Egypt promised to respect.

The military commitments of empire were both international and domestic. The imperatives towards nationalistic self-determination spoken of by President Woodrow Wilson's Fourteen Points served to challenge the very essence of the British Empire and meant that troops would need to face inwards as well as outwards. This was a problem at home as much as it was across the empire, where nationalist sentiment was gaining new voices. In London in early 1920 Field Marshal Sir Henry Wilson worked up plans to counter the prospect of 'mutiny and revolution' at home in the face of imminent strikes. He needed 18 battalions of infantry alone to guard London. Tanks were concentrated at Woolwich Arsenal and when later that year there were strikes by railwaymen, transport workers and coal miners plans were hurried to send these armoured vehicles to potential flash points across the country. Wilson was worried that he didn't have the troops needed to quell industrial disturbances in Britain and nationalist uprisings from Egypt to India and also to fight a war in any one of a number of hotspots that might emerge but for which there was no military provision. In fact, troops were so short that to manage striking workers in England Wilson was forced to order battalions from Malta and Egypt, while reserves were mobilised and a special paramilitary 'Defence Force' was created, eventually numbering 80,000.

The commitment to Russia, which had begun in the midst of war, continued beyond it into the peace because, as argued persuasively by the Secretary of State for War, Winston Churchill, the opportunity existed to defeat the Bolsheviks and thus prevent the virus of communism from gaining a further foothold in Europe. At the end of 1918 Britain had two forces in Russia, one in the far north at Murmansk and Archangel and the other in the Caucasus. The problem, it was discovered through 1919, was that the counter-revolutionary White forces were disorganised and immature, with little real chance of defeating the fearsomely motivated and organised Bolsheviks. Refusing to throw good money after bad, by September 1919 the last British soldiers were evacuated from Archangel. The arguments for staying in the Caucasus, however, were more compelling, in that a British presence would diminish the old Russian threat to India. In January 1918 a small force of 1,000 mainly Australian and New Zealand soldiers left Baghdad with the aim of taking over the line of defences that the collapsing Russians and local Armenians were holding against the Turks at Baku. General Lionel Dunsterville, the leader of this expedition, failed in this ambition, the oil city being captured on 14 September 1918 by Enver Pasha's Army of Islam. But the writing was on the wall for the Ottoman Empire and the Allies regained control of Baku on 30 October with the Ottoman surrender.

The Royal Navy's 'Caspian Squadron' flew the flag for Britain and an infantry division deployed between Batum and Tiflis (now Tbilisi), part of Lieutenant General Sir George Milne's 'Army of the Black Sea', headquartered in Constantinople. These forces were accompanied by an equally substantial force in Iraq. The subject of the military commitment to the Caucasus was to become an open war between Lord Curzon's* Foreign Office and the War Office. The former saw a military promise to vouchsafe Britain's commitment to the stability of the post-tsarist and Ottoman worlds, whereas the latter agonised every day as to where the troops to undertake these commitments would come from, given that no more money was available. Curzon plotted while Wilson fumed. Every attempt the CIGSs made to withdraw this wasteful expenditure of scarce infantry battalions in the Caucasus was countered by the Foreign Secretary. It was the inability of these troops to make a difference to the White cause which eventually precipitated the withdrawal of

---

*Curzon was Secretary of State for Foreign Affairs from 1919 to 1924.

the final three battalions in June 1920, solving Wilson's problem for him.

A similar argument raged between the War Office and the Foreign Office over the troop commitment to Persia. The country was a bastion against bolshevism, so the line of reasoning went. The country's oil was also of vital strategic interest to Britain. The argument swayed from one side to the other during 1920, only the realities of financial retrenchment forcing the issue in 1921, leading to a British military withdrawal. This left Iraq, which contained a very significant British garrison of some 60,000 troops, 50,000 of whom were Indian soldiers paid for by the British exchequer. The whole point of securing Iraq in the first place during the war had been the protection of the southern Persian oil fields. That had been done. An expensive military commitment was being undertaken in a Muslim country without the support of the local population, now for no practical purpose, apart from imperial pride. The challenge was compounded by a rebellion breaking out in June 1920. The situation was back under control by October but at a high cost in life and treasure. Indeed, quashing the uprising had been a very close-run thing. A thousand British and Indian soldiers had died in an insurgency that had brought Shia and Shiite Muslims together and very nearly humiliated Britain in the process. The slaughter of Iraqis from the air, including the punishment of villages which had rebelled, offered a new and cheap – though shocking, indiscriminate and morally insupportable – way of 'keeping the natives down'.

The new Royal Air Force (established from the Royal Flying Corps in 1918) argued successfully that future counter-insurgency problems could be solved by bombing recalcitrant tribesmen into submission. This seemed to be a useful alternative to a manpower-intensive – and potentially costly and ineffective – land-based expedition. Many of those killed in Iraq had been cut down by some of the 97 tons of bombs dropped and the 183,861 rounds fired from the air during the campaign. This lesson led Sir Hugh 'Boom' Trenchard (Chief of the Air Staff) to offer to take responsibility for the security of Iraq away from the army, with an air policing scheme in which villages could be controlled from the air. Churchill, now Secretary of State for the Colonies, put aside the moral squeamishness he had expressed in a report on such operations in 1921 (killing women and children indiscriminately from the air), for the sake of economy.[5] The government had been shocked at the huge financial cost of four months of fighting: £40 million, the equivalent of £2 billion in 2023 values. Trenchard's concept of 'air policing', supported by armoured cars and locally raised levies, would be accompanied by the establishment of a friendly Iraqi government in Baghdad, rather than the direct administration of the mandate by Britain.

———

'Home Rule', a form of independence for Ireland, was one of the longest-running domestic disputes in Britain. A political solution of sorts in the shape of the Irish politician and nationalist John Redmond's Home Rule Bill had reached its final stage during the hot summer of 1914 but was shelved by Asquith's Liberal government (which had been dependent on the support of Redmond's Irish Parliamentary Party in parliament) because of the onset of war. It was to prove one of the most egregious failures of that political generation. Threats by the Ulster Unionists to take up arms against Redmond's Bill made many in the political centre fear the prospect of civil war in Ireland. The political march to Home Rule between 1912 and 1914 led to the raising by Sir Edward Carson and Unionists in the north of an armed militia 100,000-strong – the Ulster Volunteer Force – to defend the Union from the possibility of a Catholic-dominated government of Ireland, even if that Home Rule government was designed to be largely toothless.*

In 1914 a group of officers based in the garrison at The Curragh in County Kildare, the British Army's principal garrison in Ireland, suggested that they would refuse to obey orders if they were

---

*Not to be mistaken for its later (post-1965) namesake.

instructed to oppose the Ulster Volunteer Force if the latter was determined to prevent Home Rule. In the event only resignations rather than a mutiny took place, although it demonstrated the intensity of the emotions in Ireland over the issue of a Home Rule settlement forced on a significant minority who opposed it. In the end even the resignations were rescinded, but the question remained: would the army remain loyal if Westminster went so far as to enforce a settlement on Ireland against the wishes of its loyalist (Catholic and Protestant) populations? Redmond agreed to exclude the north-eastern counties of Ireland opposed to Home Rule from the final formulation of the bill, but the excuse for shelving the entire problem was the outbreak of war.

The events of the Anglo-Irish War (the 'Irish War of Independence') which broke out in its final incarnation in April 1916 and ended in July 1921 were a trauma for both communities in Ireland, nationalist and unionist. The war was a trauma for Britain and Ireland as a whole, united politically since 1801, and for the British Army, which had long recruited heavily in Ireland and had deep roots in all communities. The wildly popular 1912 song 'It's a Long Way to Tipperary' now seemed to register the gulf between Irish nationalism and the union of the British Isles. The political leaders of both communities – and in London – had failed to secure the compromises necessary to allow all traditions in the country to co-exist in harmony.

The result was, for the British and their army in Ireland, a counter-insurgency operation that boiled over from civil disobedience in 1919 (when 18 police officers were killed and 20 wounded) to insurgency in 1920 (with 182 police and 50 soldiers killed and 387 wounded). It did not end until July 1921 when a truce was agreed, paving the way to a political settlement that was ultimately (but not immediately) to sever the majority of Ireland from Britain. For the new Irish Free State, a separate entity to Northern Ireland, which remained in the Union, it resulted in a further year of bloody civil war. For Ireland as a whole, it resulted in a hard border between the two parts, 26 counties in the south and north-west, and six counties in the north-east.

The British Army traditionally garrisoned over 20,000 troops in Ireland. In theory, their role did not include policing or internal security. Rather, Ireland was the home base of a number of infantry and cavalry regiments; following the Cardwell model, the second battalion of each of these regiments served abroad.* The role of the home-based battalion was to recruit and train soldiers, preparing them for service wherever the overseas-based battalion was posted in the empire. In the 1830s Ireland, with 30 per cent of Britain's population, provided over 40 per cent of its soldiers. By the turn of the century this proportion had dropped to 10 per cent. At the end of 1919, as part of the process of demobilising the army after the war, there were 37,259 soldiers in Ireland, with 34 infantry battalions, six of which were due to disband.

The army – distracted by the great events occurring across the channel in Flanders – was first used to restore order in Dublin following the Easter Rising in April 1916. This armed uprising in Dublin by a handful of radical nationalists (the Irish National Volunteers and Sinn Féin joining together) initially did not enjoy much popular nationalist support. Nonetheless, it was this single event that precipitated the slide into a nationalist insurgency and was to ultimately lead to a new settlement for Ireland at the end of 1921. While a full-scale military insurgency against the state didn't erupt until 1920, and then mainly in parts of Munster (primarily Cork) and Dublin, the Irish Republican Army (IRA)† campaign against the police began in early 1919. Responsibility for the maintenance of the law remained with the Royal Irish Constabulary, although at the end of 1920 primacy for operations was ceded in parts of the country – under martial law – to the

---

*The Cardwell model was designed by the Secretary of State for War, Edward Cardwell, between 1870 and 1874. In this model, recruitment into regiments was based on 66 localities across the United Kingdom. All infantry regiments were to comprise two battalions, one of which was always based at home for training purposes and the other abroad.
†The IRA were the reconstituted Irish Volunteers of the pre-war era, raised by Redmond's nationalists as a foil to Carson's Ulster Volunteer Force.

army. London didn't want to make the insurgency appear like a war, as it was believed this would confer *de facto* legitimacy to the gunmen. The insurgents were officially criminals and were dealt with as such by the police, supported where necessary by the army. The problem with this approach was that it failed to recognise the serious political purpose behind the 'war', the determination of its adherents and the deep wellsprings of support it enjoyed from across the population.

Nationalist violence presented London with a profound political dilemma. One of the principal issues with the decades-long debate over Home Rule was the seemingly unbridgeable chasm between those (Catholics and Protestants) who wanted a united, independent Ireland, and those (Catholics and Protestants) who wished Ireland to remain part of the Union settlement of 1801.* It is arguable that a chasm of this magnitude was never bridgeable, given the fundamental cultural, religious and political traditions to which the extremes of each party adhered. London didn't believe it was its job to persuade the Unionists that they had nothing to fear from a newly independent and Catholic-dominated state, but nor did the nationalists. There was a clear failure of imagination by all parties to create a settlement that enabled the majority to agree an acceptable civil *modus vivendi* for a new Ireland and for the majority party in the south to accommodate the fears and concerns of the minority. The more radical one's political position became, the less likely there appeared to be a commitment to ruling equitably for all communities. Unfortunately, most of the politics of nationalism and unionism during these years was zero sum: there was no compromise solution between the extremes offered by either side. The inevitable result was partition.

The onset of war in August 1914 had prompted a considerable display of patriotic loyalty from both communities. A very large proportion of Catholics, loyal both to the idea of independence

---

*Many nationalists were content with a form of devolution that allowed them to remain a fully functioning part of the United Kingdom.

and to the current government, joined the British Army in 1914. Most of the Ulster Volunteer Force joined up and were transferred into the British Army lock, stock and barrel as the 36th Ulster Division. It was a matter of profound regret to Redmond that the War Office didn't feel able to do the same with the Irish Volunteers. By the autumn of 1915 there were 132,454 Irishmen in the forces, 79,511 of whom were Catholics. Since the outbreak of war 81,408 had volunteered, 27,412 having been Ulster Volunteers and 27,054 Irish Volunteers.

Catholics and Protestants, unionists and nationalists, died for the same cause between 1914 and 1918. But at home they could not resolve their differences to allow themselves to imagine a new type of settlement for Ireland. For Britain, the issue was fast becoming being 'all in or all out'. If the former, a significant repression of nationalist activity – perhaps even a counter-insurgency campaign against those who wished to achieve their aims by means of bullet and bomb – would be required. If the latter, the very strong possibility was civil war. As was often the case, Westminster tried to walk the tightrope, and fell off.

The small group of radical nationalists who initiated the Easter Rising regarded Germany as Ireland's friend simply because Britain was at war with Germany. Britain's distraction in Flanders was regarded as an opportunity. The British, absorbed politically, economically and militarily with the Great War, were in no mood to play games with nationalist hardliners, especially since they had already agreed a new settlement via Home Rule which would be enacted at the end of the war. It considered this violent uprising as an attempted *coup d'état* and therefore, at a time of war with Germany, treason. Some even worried that it presaged the beginning of a Bolshevik revolution. The response was a military rather than a political one, with no eye to the consequences, especially the creation of martyrs. By means of over-reaction, the British military authorities, in suppressing the rebellion and executing the ring-leaders under the terms of martial law, served to inflame a situation that could have been handled in a far more measured way. Sixteen ringleaders were executed following shortened legal proceedings under the terms of martial law. It was a remarkably ill-considered response to the situation. With the internment of 2,000 republicans (all were released by mid-1917), the harsh response to the uprising did more than anything else to energise the nationalist call for independence and bring the uncommitted into the independence (rather than Home Rule) camp. It was the extremists who were, after the Easter Rising, now setting the agenda. London – and Dublin Castle – were too complacent or distracted to appreciate the extent of the challenge they faced from the radicals. The latter wanted independence now and were prepared to force the issue using both political and military means.

As a result of political cowardice on the one hand (the failure to execute Home Rule in 1914) and military over-reaction in 1916 on the other, leadership of the nationalist cause in Ireland slipped away from those seeking independence by political means to those who were willing to demand it by force of arms. Many 'centrist' political nationalists of the Redmond stamp who thought they had secured a deal with London, came to the conclusion that 'the English' could not be trusted. As is so often the case in these situations the hardliners, previously kept on the political periphery by the strength of the consensus, stepped into the vacuum. If politics wasn't going to work, the gun would. Once again, the army was asked to salve a problem that the political system had failed to solve.

In the period after 1916 Sinn Féin emerged, and set itself up after the elections in 1918 as the alternative in Ireland to Westminster. The British, seeing 1916 to be a law-and-order issue, never saw the strength of the new political movement which now led the nationalist cause. Westminster, focused at the time on the demands of the Great War, inadvertently exacerbated the slide towards nationalist extremism. In April 1918, with the *Kaiserschlacht* in full swing, the government announced that it was reserving to itself the right to apply conscription to Ireland, something it had hitherto foresworn. The government in the event didn't actually enact conscription, but the threat was enough, uniting even moderate opinion against Westminster. London reacted by declaring both Sinn Féin and the

Irish Volunteers proscribed organisations, arresting their leaders. The opportunity for a peaceful transition to independence had now passed and Sinn Féin went on to win a landslide victory in the December 1918 Irish Assembly elections, held under terms of the newly enacted Home Rule legislation. Thirty-four of its 62 elected representatives (from a total of 105 seats) were in prison. It then set up its own alternative, unconstitutional parliament in Dublin, the Dáil Eireann, using this electoral endorsement to boycott government by London or the Dublin Castle administration. An armed campaign against the government began soon afterwards.

During 1919 the IRA attacks were on the Royal Irish Constabulary, not merely to gain an arsenal for themselves but to remove the ability of the police to enforce the law of the land. Once armed, they moved to selected killings of police officers. During the late summer of 1920, especially in Munster Province, they progressed to groups or 'flying columns' of fighters able entirely to dominate weakly defended rural areas and, with numbers in a few cases of over 100 fighters in each column, capable of taking on the security forces in gun battles. The Royal Irish Constabulary proved incapable of countering the IRA on its own, with the result that the authorities in Dublin proceeded to recruit 8,000 additional constables, many ex-soldiers from England, the so-called 'Black and Tans' and the Royal Irish Constabulary Auxiliary Division. The approach was to 'stamp out' rebellion and restore law and order, by increasingly robust methods if necessary. However, using men little concerned with an appreciation of the political nuances of nationalism in Ireland was a serious mistake. Sir Henry Wilson saw the employment of such men as a major error of judgement and railed against the policy. The consequences of the sometimes-undisciplined behaviour of the Black and Tans tipping the population into the nationalist camp were not considered.

By the time Major Arthur Percival arrived in Cork in January 1920 (he had recently returned from the mission to Russia), the insurgency had been underway for a year. The 1st Battalion of the Essex Regiment had been sent to Ireland the previous August not because of the security situation, or because it was especially trained to undertake counter-insurgency operations, but because this was its home posting. On arrival large numbers of its wartime soldiers were discharged, and reservists and new recruits brought in, at a stroke dramatically reducing its fighting capacity. The battalion of which Percival became intelligence officer was in fact little more than a double company, comprising seven officers and 201 other ranks. Far from forming merely a peacetime garrison, taken up with training, the growing insurgency threw them into provision of military aid to the civil power, providing support to the police in the fight against a rapidly growing terrorist movement. The insurgents, of course, had home knowledge on their side, and were able to create seven 'battalions' of armed supporters in all the major towns throughout the Essex's area of responsibility. The (much reduced) battalion had no understanding of what was required nor experience of this complicated and sensitive form of soldiering. During 1920 the primary IRA activities were to steal arms and undertake individual killings of police officers. There were few large-scale actions, but a steady campaign of harassment against the constabulary lowered police morale and helped provide a sense that the gunmen, who could strike when the circumstances suited them, somehow had the upper hand.

The British Army was not prepared or equipped for the insurgency that developed through 1919. 'Trouble' – a word that was to become a euphemism for the insurgency – was very localised, and primacy remained with the police. The army in fact was entering a dangerous situation blind, unaware that it needed to work hand-in-glove with the police, that police intelligence was critical to understand who the insurgents were and where they were, and that it was the general population that would decide to whom they would offer loyalty and support. Counter-insurgency is a tough operation of war, not to be undertaken lightly or with excessive or misdirected force.

Indeed, no doctrine existed in 1919 to describe how best to take on armed insurgents hiding in plain sight in an operation where the police remained in charge. What was the best way to take the sting out of the insurgency? The primary lesson for the British Army was that in issues of internal rebellion, politics and the rule of law must remain supreme. Likewise, the police and the army need to have a united command and control system in which intelligence and action become different parts of the same approach to denying the insurgents access to weapons, public support and freedom of movement. Using violence to repress violence only reinforces the prejudices of those on the other side, or those contemplating what side to join. Violence begets violence. The outrageous use of live ammunition against protesters in Amritsar in April 1919, resulting in the killing of 379 Indians, was not merely an affront to the principle that the protection of the law extends to all citizens, even (and especially) when they are protesting publicly about an injustice, but it had very significant long-term consequences for Britain's claim to be ruling in India for *all* Indians. After Amritsar, the moral dimension to the British claim to rule justly and fairly for all was fatally undermined. Amritsar had a direct effect also on the willingness of politicians in Britain to allow soldiers to play a significant role in the insurgency campaign in Ireland.* In Ireland, the habit of turning a blind eye to reprisals for terrorist action, tacitly approved of by the Chief of Police, Lieutenant General Hugh Tudor, and supported by Lloyd George (though regarded with horror by Wilson and Macready, who considered at one point dissociating the army from the police), was entirely counter-productive and unenlightened, as it demonstrated a failure to

---

*Interestingly, the man responsible for the Amritsar massacre, Brigadier General Reginald Dyer, was, though born in India, an Irishman educated in Cork. The Lieutenant Governor of the Punjab, Michael O'Dwyer, who supported Dyer's action at Amritsar and cracked down on protests after the event, was a Roman Catholic from Co. Tipperary. History is complicated. See Nick Lloyd, *The Amritsar Massacre: The Untold Story of One Fateful Day* (London: I.B. Taurus, 2011).

understand that harsh security force behaviour, especially illegal reprisals, alienates the very population the security forces need to win over.

The IRA's attacks had first been against barracks and police stations, but these had, on the whole, been failures. The strategy was changed to target individual Royal Irish Constabulary officers, usually when they were at their most vulnerable, such as at home or undertaking social activities outside of work. As a result, police morale quickly dropped and the task of providing policing in rural communities became a challenge, as the insurgents had hoped. Their ambition was to cause Westminster's rule of law to be impossible.

The security forces' first inclination to dealing with the threat of trouble was to rush troops to the affected area in trucks, to try to round up the insurgents. Vast and complicated sweeps were undertaken through rural areas looking for 'the enemy'. It was a naive and ineffectual stratagem. Most such attempts were fruitless. The very nature of insurgency is that the gunman can hide in plain sight, hidden within a supportive or cowed population and only come out to fight when the advantage is theirs. Road-bound infantry patrols also became easy targets, the insurgents mounting a series of skilful ambushes during late 1920 and 1921, making the most of the advantage of surprise. Big sweeps, curfews, public searches and restrictions of movement served not to find any culprits, but merely to irritate the very people whom the army needed to have on their side.

As the months went by Percival realised that with no intelligence about who the insurgents were, where they lived and how they operated, all the military force in the world could only react against insurgency, rather than pre-empting or countering it. It was in the careful gathering and sifting of information that Percival and his men, over the months that followed, became the IRA's most feared opponent in Cork. Using the tactics of the insurgent to fight him at his own game, rather than deploying the tactics of conventional warfare, gave the greatest results. Ambushes, working at night, camouflage and military clothing adapted to the circumstances (such as rubber-soled boots), unannounced patrols, plain clothes intelligence gathering through the police and careful observation of suspects, their homes, places of employment, patterns of life and daily activity, as well as searches for weapons and individual insurgents based on this intelligence, became foundational approaches to Counter Insurgency (COIN) operations. Using intelligence from networks of informants to capture or kill terrorists while at the same time separating them from their sources of support and sustenance was all important. The capture by Percival in July 1920 of both the leader and the quartermaster of the IRA's West Cork Brigade, Tom Hales and Patrick Hart, using these tactics, was a significant coup and vindicated an intelligence-led policy.

As violence gradually increased, so too did the numbers of security personnel. During the 18 months between January 1920 and the truce of 11 July 1921, 525 police and soldiers were killed and 935 injured. A full three-quarters of these occurred in the final half of the period. Likewise, during the final six months there were about 1,460 civilian casualties, of whom 707 were killed. In reflecting on the casualties of the Anglo-Irish civil war it is important to note that the majority of so-called 'British' casualties were in fact Irishmen, who died serving in either police or army, fighting fellow Irishmen. It was a civil war in most respects. By the end of 1920, police numbers totalled 10,000 with 50,000 troops. By the time the truce was secured the army had poured 60,000 soldiers into Ireland. In the year leading up to the truce it was clear that Westminster had lost the south. IRA violence increased. Sinn Féin refused to co-operate with the civil administration from Dublin Castle, continuing to undertake its own governance. The elections in May 1921, following from the Government of Ireland Act, which established separate parliaments for the six counties of the north and the 26 of the south, showed where the overwhelming political sympathies of the constituents of both parts of Ireland lay. In the 26 counties only four from 128 seats went to Unionists. In the six counties it was 12 seats to Sinn Féin from 52 seats. An upsurge of violence by the IRA following the elections, together with the continued refusal of Sinn Féin to govern under the rule of Westminster, meant that the only way the 26 counties could be governed was by martial law. There was talk of a permanent garrison on counter-insurgency duties comprising 250,000 men, with no certainty of Sinn Féin's revolt ending. As this was not acceptable to London, Lloyd George offered Sinn Féin a truce followed by the promise of talks on a future devolved settlement. This led to the signing of the Anglo-Irish Treaty in December and the establishment of the Irish Free State in the 26 counties. The six counties in the north would remain unchanged within the union of 1801. Major Arthur Percival packed his bags and, with the rest of the British Army in Ireland, travelled north to a new home in Carrickfergus. The Anglo-Irish War was over, with a settlement (a divided island) that neither wanted but which the insurgency had made impossible to avoid. It had stretched the British Army, giving it tasks for which it was unprepared and untrained. Was this type of counter-insurgency campaigning to be its bread and butter, distasteful as it was, into the future?

Chapter 7

Imperial policeman versus continental commitment

It wasn't just troop numbers that mattered in the post-war army, but pounds, shillings and pence. A peace dividend was expected after the crippling cost of the war. In addition to the dramatic slashing of defence expenditure following the end of the war, a moratorium in August 1919 – the suggestion of Winston Churchill – was placed on any further spending under the so-called 'ten-year rule'. This was a Treasury assumption that there wouldn't be another major war within ten years, thus keeping the lid on expenditure:

> It should be assumed that the British Empire will not be engaged in any great war during the next ten years, and that no Expeditionary Force is required for this purpose ... The principal function of the Military and Air Forces is to provide garrisons for India, Egypt, the new mandated territory and all territories (other than self-governing) under British control, as well as to provide the necessary support to the civil power at home ...[1]

There was no money for military expansion, but neither was there any for innovation, combined-arms training or preparation for future expeditionary warfare. In the two decades that followed it was the crippling freeze on expenditure that did more than anything else to constrain the army's modernisation and its preparation for modern war. The role of the army was to guard the empire. So far as His Majesty's Government was concerned, the war to end all wars had done its job. There was no need to consider or plan for a future one, whether in policy, financial or practical terms. One of the consequences of this moratorium was its destructive impact on Britain's domestic arms industry. Another was the fact that it made everything an issue of money, giving the Treasury the whip hand in considerations of national defence. Another was the disincentive it gave to the War Office generally to undertake forward strategic planning (or doctrinal development), at least insofar as another 'war of the highest magnitude' in Europe was concerned. The rule was to remain in place, ruinously, until 1932, after which it was converted to a five-year rule. War became progressively hypothetical. Why was the British Army so unprepared for war in 1939? The principal reason was that the British government, by means of its faulty defence planning and financing in the previous two decades, made it so. It was not the first time British governments have entered periods of war unprepared but this time they had done so despite the warning bells of a deteriorating security situation in Europe and the Far East ringing alarmingly for several years beforehand.

Beyond this, in the face of steadily increasing government expenditure and rising taxes, was an attempt from 1922 to reduce government spending overall. This was an all-party affair. None of the Conservative, Liberal or Labour parties was keen to see expenditure on defence. The committee, chaired by Sir Eric Geddes, reduced total defence expenditure from £189.5 million in 1921 to £111 million in 1923, of which £43.5 million was for the army. The Geddes 'axe' was an act of grievous self-harm. This zero-sum approach (i.e. within the limited budgets available, where one party gained it was at the expense of another) was a devastatingly foolish way to manage something as serious as the defence of the national interest. There were few incentives to co-operate between the services and fewer still to collaborate, because at the heart of the relationship was always a struggle for money. The exercise was undertaken by Treasury civil servants with no military advice, and who made a series of egregious errors in their assumptions. One of these was that modern weaponry (such as machine guns and tanks) made fewer troops on the battlefield necessary. The army's annual budget fell steadily every year thereafter to £36 million in 1932. Current commitments aside, it was no way to prepare for another war. In any case, so far as Geddes was concerned, no *real* war was predicted. The future security of the country was, to his committee, an expensive and bureaucratic inconvenience. Wilson was horrified at Geddes' approach. In a fierce rear-guard action towards the end of his time as CIGS he complained that Geddes didn't understand the world in which Britain was now operating, and that his report 'had been careful to avoid any discussion of the purpose for which the British Army exists'. The report 'not only reveals a complete misconception of the reasons for the Army's existence, but ignores the responsibilities with which it is entrusted'. Ten pages of detailed criticism of Geddes' assumptions followed, emphasising the scale and extent of Britain's overseas commitments:

For the last hundred years the British Regular Army has been maintained, not for intervention in a European war, but the protection of our overseas territories, for the support of the Civil Power in the maintenance of law and order throughout British territory at home and abroad. The aim of the Army has been regulated by our overseas commitments, and not the threat of any Continental Power. It is true that as the German menace became more and more imminent in the years succeeding the South African war, the possibility of our being involved in a European campaign had to be increasingly taken into account. Consequently, those military units that we were obliged to provide at home for the maintenance, relief and support of our overseas garrisons were gradually organized in such a manner as to allow them, such as they were, to intervene in support of our Allies, should His Majesty's Government decide so to act. But the size of the Expeditionary Force had no relation whatever to the strategical problem of a Franco-German conflict. This force consisted merely of the spare parts of our overseas military machine, assembled to form a provisional weapon with which to meet the threatening emergency. Thus, the disappearance of the German menace in no way affects the size of the British Army, which is conditioned by our overseas commitments.[2]

The Geddes Committee recommended a reduction in the establishment of the British Army to 130 infantry battalions and 19 cavalry regiments for all purposes. Wilson, notwithstanding the soundness of his arguments about how few troops were at the time available to undertake the considerable increase of tasks placed on the army since 1914, allocated them as follows:

Station	Cavalry regiments	Infantry battalions
Home (including Germany)	8	67
Abroad (see below)	11	63
India	8	45
Aden	0	1
Egypt	3	7
Constantinople	0	2
Singapore	0	1
Hong Kong	0	1
North China	0	1
West Indies	0	1
Malta	0	3
Gibraltar	0	1

The 45 British battalions in India constituted 60,000 men, supporting an Indian Army of 190,000 (100 infantry battalions, 21 cavalry regiments and 18 mountain artillery batteries) formed into 20 infantry regiments in 1922.

An unfortunate side effect of Wilson's vigorous defence of the size of the army commensurate with its worldwide garrison commitments, was that it helped reinforce the idea that the army's sole task was to be an imperial gendarmerie. Indeed, the underlying assumption at the heart of British defence policy through the

1920s was that Britain would not have any commitment into the future to provide troops to the continent and certainly not to fight the type of war that had just ended. Indeed, anything else, such as an expeditionary force for despatch to a global trouble spot, continental or otherwise, was explicitly excluded from his response to Geddes' proposals. Wilson accepted the assumption that there would be no major war for ten years and used this to ignore the need to prepare the army to fight a war of the greatest magnitude. Under his watch the British Army fell asleep at the wheel. The moratorium on not planning for a war during the coming decade was taken as the excuse for not preparing – in thinking, doctrinal or organisational terms – for a 'war of the greatest magnitude' in Europe or anywhere else.

Wilson had also misunderstood the politics of the times. No one was listening to him. According to Geddes the era of big wars was over. The army had been demobilised, the population traumatised and depleted by sacrifice. There would be no more continental adventures, where Britain would be inexcusably tied, so the argument went, to the discredited strategic follies of the French, their treasonous armies and villainous generals. The empire could be defended by many fewer troops than had been necessary in 1914. Indeed, a small professional army focused not on any continental commitment (which Britain had anyway eschewed prior to 1914) but on the needs of empire sat perfectly with the traditional approach of relying on the Royal Navy, reinforced after 1918 with the Royal Air Force, for the protection of Britain's global interests. Blaming it for the human losses of the Great War, failing to recognise the role that it could play in preventing another, the country and the people turned on the army and did the demolition job the Germans had so inconveniently failed to do during the recent brouhaha in Flanders. The Australian journalist Alan Moorehead described the mood change:

> Disarmament governed all. Regular officers were being 'axed' right and left and many felt themselves lucky if they got out with a pension. The hectic backwash into peace had even gone to the

point where army officers returning from France were heckled in London streets, called warmongers; and in a world of growing hunger and unemployment, they were accused of hanging on to safe and idle jobs at the public expense. It was the final swing of the pendulum from the days when hysterical women went about offering white feathers to civilians. Gloom and lethargy pervaded the Army.[3]

The slaughter of the trenches was bad enough. For many people these losses had become a human tragedy, brought about by military incompetence. The army went into the 1920s with an increasingly tarnished reputation. War wasn't something that could be left to the generals. Even the wartime prime minister got into the act. Lloyd George blamed the generals for the slaughter, his animus for Haig in particular being clear to see. The war was, he asserted in his *War Memoirs*, a story:

> of the two or three individuals who would rather the million perish than that they as leaders should own – even to themselves – that they were blunderers. Hence the immortal renown and ghastly notoriety of the Verdun, Somme and Passchendaele battlefields ... the notoriety attained by a narrow and stubborn egotism, unsurpassed among the records of disaster wrought by human complacency.[4]

In June 1918 Duff Cooper overheard a fellow regimental officer comment that 'I shall be glad when this damned war is over and we can get back to real soldiering again.'[5] This wasn't surprising as most senior officers of the era had cut their teeth in one way or another with the army abroad. The sense that 'the war to end all wars' was an aberration was commonly expressed. Major General J.F.C. 'Boney' Fuller claimed, looking back from the vantage point of 1935, that this was the desire of 90 per cent of military men.[6] The desire among many was to revert to the comforting certainties, structures and approaches of a time long before the war clouds arrived over Europe in 1914. The Great War would never be repeated. Accordingly, by 1919 everything had reassumed its old shape. The war had been an irregularity. Now that peace had been achieved, the rush to recreate the halcyon military days of 1914 was pell-mell. Writing 16 years later, Fuller's anger at the naivety of this worldview – and especially one that failed to place *his* tank at the centre of the new military world – reverberates across the page. A black hole of unreason and exhaustion, he argued, swallowed up the entirety of the British Army's thinking class. 'Were these lessons and the thousand and one others which the War drilled into us at a cost of one million lives and eight thousand millions in money learned or even examined by the Higher Command?' Fuller asked. 'No' was his rhetorical response. The British Army, in its rush to put the Great War behind it, reverting to the red coats and burnished helmets of yore, thus 'blindly stepped backwards and fell over the precipice of long vanished hopes'.[7] Fuller's despair was tinged by bias; in the new, post-war world, his beloved tanks were forced to play second fiddle to the infantry.[*] And he forgot that it wasn't the army that wanted to reduce its size and scale, but the country. But he was right in one respect. There was no putting the clock back.

The relief of 1918 that the world could reset its clocks to 1914 was not just a military fantasy, but a social and political one too. There was a sense that Britain – and the world for that matter – had just been through an extended nightmare which, now behind them, could be consigned to history as a deviation from the norm. This was, on reflection, unsurprising. The *mentality* of Britain, and the men who led its army, had not changed as a consequence of the war. The army was, and had been for a hundred years since Wellington, a *garrison* and an *expeditionary* army, able to provide sufficient well-disciplined troops in scarlet or khaki primarily to defend the overseas empire, with a small contingent at home to provide a recruitment and training base and, when needed, military

---

[*]Fuller's theories, including 'Plan 1919', are discussed in more detail in Chapter 10.

support to the civil power. The people of the United Kingdom saw no need for anything more than this. The move after 1906 to create an army able to make a continental commitment in the event of war in Europe was itself an aberration, brought about by the particular circumstances of German militarism, now defeated. On the high seas the Royal Navy was not matched by any other country and was the primary guarantor of the country's interests abroad. During these 100 years no threat had presented itself to the British Isles. The Crimean War took place far from Britain's shores and challenged neither home nor imperial hegemony. The British Army, supported often by locally enlisted troops, was involved only in small wars in hot places, all of them far from home, such as Africa or Afghanistan, where occasional defeats could be blanketed in a popular narrative of adventure, manhood and patriotism. These wars never challenged or changed the political status quo. In any case, the Royal Navy was there to prevent boarders if anyone was foolish enough to cross the English Channel.

The end of the Great War would mean, it was hoped, that the British Army could reset itself into its traditional tasks and obligations. The vast inflow of talent and skill that had so rapidly and successfully transformed the army into the great machine that had conquered in 1918, as quickly departed when the need ended. Britain's relationship with its army reverted to its ancient norm, as the temporary amalgamation of country and army came to an end with the advent of peace. But as the war in Western Europe stopped it became clear that the world ushered in by the New Year of 1919 was already very different to that which had gone before. Indeed, one outcome was that an even more fractious world had emerged from the ruins of the old. New commitments were added to the old ones which were to stretch the British Army far beyond even the busy days of the late Victorian era. British troops remained in Russia. Major parts of the old Ottoman Empire found themselves, after Versailles, under British mandate and demanding a significant British military commitment. That in Palestine, largely unwanted, was to drain scarce troops later in the decade and in the 1930s would pitch Britain into a resource-intensive and unwinnable confrontation with both Arabs and Jews. Likewise, the British Army had to find 45,000 troops for the Rhineland to enforce the Versailles settlement. At the same time, historic animosities continued, in Somaliland, Afghanistan and Waziristan and garrisons continued to be required for India, Ireland, Cyprus and Egypt.

The operations which absorbed the energies, interest and attention of the empire after the demise of the Great War were those which the Field Service Regulations described as 'warfare against an uncivilized enemy'. Field Marshal Sir Henry Wilson made this point bluntly in January 1922, namely that 'the main business of the Army in time of peace is to act as an Imperial police force throughout our Eastern Dominions'.[8] This busy-ness meant that there was to be categorically no role for the army for another war of the like that had just passed in Europe. The principle by which the empire was to be defended was based on manpower, rather than technology or mechanisation. It meant that colonial troops could also be co-opted into the defence of the empire, but it also meant that imperatives to development and innovation – such as the mechanisation of military transport and the introduction of engines to replaces the horse of *both* the cavalry and the baggage train – received significant opposition in the years ahead. This opposition, to give its proponents the benefit of the doubt, wasn't because people were naturally resistant to change but because these innovations, in the context of the defence priorities of the time, were not considered necessary. In other words, the British Army was constructed around the idea of empire: the BEF of 1914–18 was an aberration. This was a reversion to type, as it was the nature of fighting that the British Army had as part of its DNA, gifted to it as a consequence of Victoria's wars. The 1920s saw a reversion less to 1914 than to 1890.

The 18-year-old Hastings Ismay had passed out of Sandhurst in the top 30 cadets in 1905 and soon afterwards joined his cavalry regiment – 21st Prince Albert's Own Cavalry (Daly's Horse) Frontier Force – in Risalpur on the North West Frontier. The task of the 21st Cavalry – the defence of empire – sat at the heart of

Britain's army, together with that of its cousin, the Indian Army. The North West Frontier reflected this focus, with two threats. The first was to internal stability and order, the second was a threat to its borders and the ultimate sovereignty of the colony. This was, until the Great War inveigled its way into Britain's national concerns, the primary responsibility of the British and Indian Armies.

In the post-war era a new internal challenge to the army had also arisen, in the form of the Royal Air Force, which aspired to move into the peacekeeping business. To his professional irritation – he tried hard to engineer a posting to France – Ismay was destined to spend the whole of the Great War in the 'British Somaliland Protectorate' holding Mohammed Abdulla Hassan's jihad at bay. The war halted all but sporadic forays against the dervishes. With the war over, the British returned to thinking about ways to remove the problem of Hassan – the 'Mad Mullah' – without going to the expense and trouble of another one of the pre-war punitive expeditions into Dervish territory which had never in fact done much good. In accordance with the new War Office policy of using the Royal Air Force to put down insurgencies from the air, six DH9 aircraft, each of which could carry 460 pounds of bombs, duly set off on 21 January 1920. The operation was not a success. As Ismay relates, one aircraft 'had a forced landing, four failed to locate the target, and the attack was delivered by a solitary machine, whose pilot had got separated from his companions and happened to spot the objective. Extraordinary to relate, the first bomb singed the Mullah's clothing and killed his uncle, who was standing next to him.'[9] Thereafter the Mullah and his followers retreated to deep caves and three days of bombing availed nothing. The Dervish rebellion was eventually put down by precisely the means London was reluctant to countenance in the first instance: boots on the ground.

Notwithstanding the reality of what had actually happened, the campaign gave the new Air Ministry the opportunity to argue in Whitehall the efficacy of independent air action to achieve military and political aims across the empire. According to Ismay, they argued, to no serious counter from those informed of the true facts, that a mere six aircraft had in 21 days solved the problem which had foiled the army for 21 years. The Air Ministry also, with no evidence, argued that the bombing had demoralised the dervishes and this had contributed to their defeat. Bombing had, they argued, a distinguishing effect on the willingness of an enemy to continue to fight. In Sir Henry Wilson's 1922 riposte to the Geddes Committee, he observed that 'it appears to be the advent of the Air Force which has influenced them in this direction.' He acknowledged that aircraft could undertake long-range reconnaissance, noting that it was this argument which had already removed four cavalry regiments from the army. 'But to assert that the place of cavalry can entirely be taken by aircraft in the work of close reconnaissance protection and support is a complete fallacy.' Aircraft were good, he acknowledged, in a clear dig at the moral paucity of this policy, 'for bombing women and children' but could not replicate roles that could only be undertaken by ground troops.[10] It was to prove an important moment in the debate in the next two decades about the efficacy of air power.

As if to confirm that the old world was reasserting itself, in February 1919 Habibullah, the Emir of Afghanistan, was murdered. In the confusion and turmoil that ensued, nationalist feeling was whipped up in the Punjab, followed by the crossing of the frontier by Afghan troops on 4 May. This was repelled by the 4th Indian Division. A second incursion in the Kurram Valley was repelled by the 16th Division under the command of Brigadier General Reginald Dyer, a man who had egregiously put down a civil disturbance at Amritsar on 13 April with live rounds, killing at least 379 people and wounding 1,500.

The trouble in 1919 firmly re-established the North West Frontier in the imagination of British military responsibilities. This was what imperial defence was all about. Permanent garrisons were established at Razmak and Wana to protect the country from further Mohmand raids, and to keep Waziristan in order. In 1935 a two-brigade expedition was launched against the Mohmands of Afghanistan and in 1936 the Fakir of Ipi was behind a rebellion that required 30,000 troops to put down. Jack Masters described the outbreak of violence in 1936 at Biche Kaskai in Waziristan, 'where Waziri tribesmen ambushed the Bannu Brigade on a carefully laid, well-concealed, and boldly executed plan ... The brigade suffered 130 casualties and lost many arms and much ammunition. The tribesmen, elated by this early success, went on to higher things' and operations against the Mahsuds in the south and the Wazirs in the north continued through to 1939.[11] But the truth was that the North West Frontier was always a sideshow, an excuse for inactivity and prevarication elsewhere.

Overwhelmed by these numerical responsibilities after 1919, exhausted by the political battles to retain numbers at a sufficient level to deal with brush fire contingencies, the army was relieved to be back with the thrilling routine of Kipling's wars rather than the relentlessly grim industrial slaughter of the Western Front. In the years which followed 1918, it showed a distinct lack of interest in developing the techniques and tactics associated with high-intensity battle. This was, perhaps, understandable, given the dramatic threat to the management of commitments that occupied successive CIGS and the War Office and the default setting of the army, at least until 1934, that it was an imperial gendarmerie. The Great War had been intellectually and morally *exhausting*, as well as appallingly destructive and expensive in blood and treasure. It is no surprise that the army wanted a rest from the utter relentlessness of thinking about, and engaging in, mass warfare.

A consequence of slipping back into the comfortable certainties of the army's traditional imperial role was that very quickly Britain lost a sense of what the army *could* do, and what, politically, it *should* do on behalf of the current and future security needs of the country. Being an imperial policeman was a functional, utilitarian task. There was a job to do, and a simple one at that, requiring an infantry-dominated force posture with a relatively simple technological requirement in terms of weapons and equipment. But in reverting to utilitarianism the army lost a sense that it should also be a 'force in being' (a naval concept) for the possibility – remote no doubt – of once again fighting a high-intensity war in Europe. It had been, after all, the fundamental grounding of Royal Navy policy for centuries: be ready for anything and be able to fight and beat the biggest dog in the yard if called upon to do so. By constructing an army able only to deal with the issues of the moment, the country was denuding itself of the opportunity of dealing with the problems of the future. A hollowed out army was no longer able to offer itself as a deterrent to future bad behaviour in Europe, nor was it able to offer itself as a coalition partner to a country such as France when it might be required. In short, the problem with reverting to an imperial gendarmerie was that Britain was returning to a view of the army and of war through a distinctly Victorian lens. But that age had passed. Industrial warfare had burst onto the European scene and henceforth would require a post-1918 military response rather than a pre-1916 one. The problem was that few in Westminster – either in the War Office in Whitehall or in No 10 or No 11 Downing Street – were able to discern the implications of this change. Certainly, there were no politicians willing to expend political capital on creating an army for the future. And there was no one in the War Office during the tenure of successive CIGS, good as they were at managing the problems of the day, who saw the need to think about the future with intensive clarity or, if they saw the need, were able to mobilise their respective Secretaries of State for War to think ahead. During the inter-war period in the War Office, the future had no champion.

The simple process for 'future proofing' the army would have been to ask the question: 'To the best of our knowledge, *how* will warfighting be undertaken in twenty, thirty or forty years' time?' With the answer to this question a reasonable attempt could then have been made to create force structures, equipment and doctrine, together with recruiting and training to deliver these imperatives, to ensure that when this point in the future was reached, it was accompanied by an army able adequately to respond to these security challenges. There was no attempt by the War Office in the 1920s or 1930s to answer this question. Instead, most thinking was undertaken on the basis of one or either of two propositions. The first was to provide for the current demands of military commitments. This was the immediate response of Westminster following the end of the Great War. The second, which was the approach of Geddes and the Treasury thereafter, was simply to cap the resources available, and cut the army's cloth accordingly. Neither took cognisance of future liabilities, and it was in this area that both Downing Street and Whitehall fundamentally failed the people of Britain in the inter-war period. If the army is anything, it needs to be not merely the utilitarian tool of the present, but the guarantor of utility into the future, across the full extent of security possibilities which might be required not just today, but tomorrow. In this regard, the thinking and the failures of 1923 have a worrying resonance for modern times.

The victories of 1918 had demonstrated that Clausewitz's injunction had been right: military power on land was an important instrument of polity, not merely one of incidental utility. By 1916 the Great War was in danger not of achieving the political objectives of any one country, but merely of draining each side's army of blood and the exchequers of treasure. But by 1918 the British Army had come up with a method of warfighting that meant that military effort could be focused directly on winning battlefield victory, and thus achieving the ends of national polity: namely winning the war and successfully defeating German aggrandisement in Europe. This focus on using an army to achieve a political result (rather than in 'just' fighting battles) was demonstrated perfectly by the German Wehrmacht in Poland in 1939, which introduced the world to a new word – '*Blitzkrieg*' – lightning war. To the astonishment of the world, it repeated this triumph during the short ten-week campaign in France in 1940. If Britain's government had forgotten its Clausewitz, its army had forgotten the formula of successful warfighting discovered in 1918. Arguably, it had lost it by 1919, distracted by the multitudinous pressures of peace including a pacific electorate, a parsimonious Treasury and trouble in Ireland, let alone across the empire.

Chapter 8

The modern major general: more categorical or allegorical?

During the Hundred Days between August and November 1918 the British Army found the formula for operational and tactical success in mobile, combined-arms warfare that broke the deadlock of the First World War. The problem was that it forgot the formula just as quickly. After 1919 little was done to preserve in the army's corporate memory the detailed knowledge and understanding of *how* victory was achieved. Because no one deemed it important to retain the knowledge of this type of warfighting or recognised its importance as the basis of all-arms doctrine, the debates of the 1920s and 1930s were to prove to be rudderless. The army ended up in stormy seas, tossed about by a mixture of political direction that rejected the idea of retaining an eye on the possibility of fighting another 'war of the greatest magnitude' against a peer opponent in Europe; under pressure to find sufficient troops to fight the small wars of empire (some of which, as has been seen, were very close to home, and hardly 'small'); and with no clear doctrinal framework about the ultimate purpose of mechanisation and armoured warfare.

It is extraordinary to relate, but the British amnesia as to why victory had been achieved in 1918 was profound. The authoritative *Britain's Modern Army* published by the British Ministry of Information in 1941 asserted that the Great War ended because the

The Allied offensives, 1918

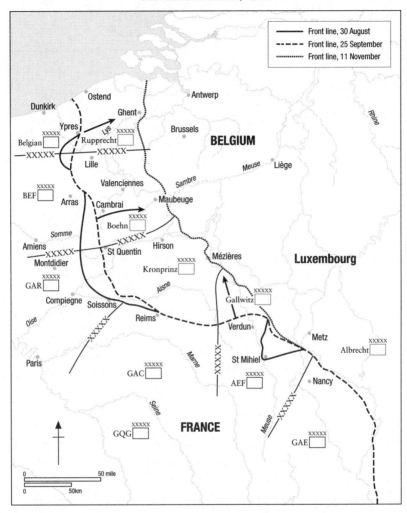

Front line, 30 August
Front line, 25 September
Front line, 11 November

Allies 'had worn down the vanquished by a combination of means, many of them other than military, [rather] than in consequence of any spectacular defeat on the classic lines of Waterloo or Sedan.'[1] This is nonsense. This publication, reflecting widespread ignorance about how the final stage of the Great War was fought, gave no recognition to the reality that by new methods in warfare and a new approach to battle, the Germans had been undone. Extraordinarily, it seems the British simply forgot the role of the battlefield in bringing about victory, falling back on a widely held prejudice that it was materiel exhaustion alone that had brought about Germany's demise. This prejudice was combined with the slander of incompetence against the generals. It was popularly held that it was not possible for dunderheads to have secured a great victory, so it must be ascribed elsewhere. Thus, by such claims were the successes of the Hundred Days dismissed as being inconsequential – or secondary – to victory. The British Army thus entered the Second World War blind, without any unequivocal view as to how the last one had ended and therefore oblivious to the pattern that the new one would take. Most decision-makers assumed the war would start as the previous one had, and follow the same shape, of two massed infantry-based armies attempting to secure an advantage over the other in an encounter battle, the conflict resolving itself in terrifying linearity. Although that was explicitly what 1918 had demonstrated would not happen, no one, in Britain at least, seemed to understand that linearity had been defeated in 1918, and mobility – even manoeuvre – re-established on the battlefield.

Justifying why it was that Britain and France found themselves unprepared for *Blitzkrieg* in 1940, the author or authors of this 1941 British Ministry of Information book tendentiously claimed: 'A repetition of the war of 1914–18 might reasonably have been expected to display this same character of a long drawn-out, slow-moving drama, braked in its tempo and clogged in its development by an over-balance in favour of the defence.'[2] But this was an egregious misrepresentation of the state of affairs in France for most of 1918. The Great War did *not* turn out like this, which makes the claim 'might reasonably have been expected' an admission of astonishing ignorance. By such assertions can the failures of the post-Great War period to develop a warfighting doctrine, based on military victory in 1918, be understood. The events in France in 1940 were to demonstrate unequivocally – but too late – the vacuity of these assertions. The truth was that Britain and its army were entirely unprepared for the real innovations in warfare in 1939 and 1940. These were improvements extensively adopted by the BEF in 1918 but subsequently forgotten in Britain's collective memory. They had not been forgotten in Germany. These were not merely about the tank or the bomber. The innovation of 1940 was about an approach to warfighting in which a fast-moving and powerful army found and exploited vulnerabilities in the enemy's defences, and by means of speed and aggressive action out-thought the enemy's decision-making cycle. It was about the Hundred Days again, on steroids.

This particular 1941 publication claimed that 'at Cambrai [in November 1917], and later during the final Allied advance of 1918 in the west, the tank led the way at every stage and proved itself to be the tool of victory'. This was a grievous overstatement, an argument that could well have been made by J.F.C. Fuller himself. It represented what a significant minority of British military thinkers and practitioners believed. In truth, however, while the tank was a key element of the solution in 1918 as in 1940, it was only one part of it. To ascribe sole agency in victory to the tank is to seriously over-rate the role of this innovation. As Duff Cooper's experience had shown, breakthrough had been achieved as the result of a close working of a combination of factors. The tank on its own was not decisive in victory in 1918, although working closely with accurate artillery and new tactics for the infantry it proved unanswerable by the Germans on the Hindenburg Line.

The 1941 publication *Britain's Modern Army* proceeded to describe the tactics of *Blitzkrieg*, expressing some surprise at the ability of the German Army to innovate so spectacularly. It entirely failed to appreciate that these tactics were the inevitable development of those that had been so expertly deployed by the British Army in France in 1918.* The book goes on to explain how these offensive tactics could be countered. The answer, it asserted, was defence in depth, the book arguing that the 'principle of a wall in front of the zone with the idea of preventing the enemy

---

*Incidentally, tactics also deployed by General Allenby in his September offensive in Palestine that same year.

setting foot inside it must be discarded'.[3] Quite. But as we have seen this idea had also been discovered in both 1917 and 1918, and as a tactical doctrine had been well explained in the 1920 edition of the Field Service Regulations (although it was much less clear when considering the subject from an operational perspective). It should not have been a new lesson in 1940. The simple truth is that in Britain at least the lessons of 1918 had not been learned. The British Army had to re-learn them the hard way. Failure occurred in 1940 therefore because of the inability of the post-war army to take seriously the need to capture the experience of 1918, and convert it into a modern warfighting methodology. Germany had not made the same mistake.

When General Bernard Montgomery led the 21st Army Group through France and into Germany in 1944 and 1945 he deployed precisely the same methodical approach to warfare that he had seen work so successfully in 1918. Indeed, his plan for El Alamein in late 1942 followed essentially the pattern of 1918, at least insofar as the advance against the Afrika Korps was fundamentally methodical. Clearly, it had moved on since 1918, but the principles remained the same: fully integrated armour and infantry advances against weak spots in the enemy defence, carefully orchestrated with artillery bombardments and fire support both on land and in the air, co-ordinated by radio. But in the immediate aftermath of the Great War many influential soldiers did not consider that Amiens presaged a new way of fighting. In a lecture to the Royal United Services Institute in November 1919, the first anniversary of the Armistice, Major General Sir Louis Jackson, for example, saw no future for the tank. 'The tank proper was a freak,' he opined. 'The circumstances which called it into existence were exceptional and are not likely to occur again. If they do, they can be dealt with by other means.'[4]

It was only in 1932 that the War Office belatedly ordered a review of the *fighting* lessons of the Great War. A committee was formed under Lieutenant General Walter Kirke and the findings were published in October 1932.⁵ On the whole, it is an excellent document. It is very short (42 pages but accompanied by appendices examining two assessments of the Western Front, and one each on Gallipoli and Egypt/Palestine) and catalogues a range of useful lessons encompassing military strategic, operational and administrative issues. The four sections – Peace Preparations, Strategy and Tactics, Equipment, and Organization and Training – capture a range of useful lessons. Surprisingly, however, the two appendices on the Western Front concentrated their analysis on the first two years of the war, rather than the final two. There was no systematic analysis of 1918 nor of the Hundred Days, the subject that was most needed in terms of attempting to understand how the next war should be fought. Critically, and perhaps because the Kirke Committee's findings were biased towards the start of the war rather than its end (the writing of the Official History had by that time only reached 1916), no new doctrine about operational methods of warfighting resulted. Kirke was under no illusions about the Big Question, however. 'As the attack is by far the most difficult operation of modern war,' he noted, 'our small army should be able to strike quickly and effectively, and therefore a full measure of reorganisation and re-equipment should be directed towards its solution.' Without the evidence to hand of the experience of 1918, Kirke and his committee were looking through a glass darkly. Nevertheless, they perceived a number of issues that were to play a significant role in the next war, including the need for radio communications, the direct support of infantry battalions by close support artillery and the intimate working together of infantry, mobile artillery and armour. On the subject of armour, the report concluded, there 'was not an adequate reason for delay' in expanding the tank arm. One cannot help thinking that President Zelensky's Chief of the Ukrainian General Staff might have come to the same conclusion in 2023.

Unfortunately, nothing much came of the report. There was simply no political imperative at the time to spend any money on bringing the army up to a state where it could fight another war in Europe. Most soldiers saw this and didn't challenge the overwhelming view that the next war (if it occurred) could be won from the air. An attempt was made by the War Office in 1934 to summarise some of the points emphasised in the Kirke Report for the benefit of the army. The points made were somewhat general (and included such things as the importance of surprise in the attack and the need for commanders to adopt the 'indirect approach'*), and the effort otherwise fell flat. The War Office's specific failure was not to order its thinking in definitive terms about how the British Army needed to fight the wars of the future. But it was clear that history (the Hundred Days in particular) had already been forgotten. If it had not been so disregarded, the decisive lesson of 1918 – that success in battle was based on a careful blend of firepower and mobility, with tanks, armoured and mobile infantry, armoured and mobile artillery, effective anti-tank guns and tactical aircraft working in a combined manner both in attack and defence – would not have had to be learned the hard way again in the years from 1940 to 1945.

If the army did not simply understand what it had just experienced, it is true also that it did not see the need to think about or materially prepare for the type of high-intensity *warfighting* it had just engaged in – successfully – in Flanders. This was because the entire practical, financial and intellectual commitment of the British Army was now focused, or re-focused, on the empire. High-intensity combat was seen to be exclusively a problem for war on the continent, and such a commitment was anathema to the new post-war Britain and would be for the next two decades. 'In war we were proud of our new weapons,' lamented Fuller 'but in peace we anathematised them'.[6] Instead, the army ignored the new weapon systems and proven methodologies of 1918, rushing determinedly back to the certainties of the pre-war period, an era dominated by infantry and horsed cavalry. Britain and its army seemed determined to stick its head in the sand. Thanks to the national hysteria that flatly denied the possibility of another war in

---

*The influence of Captain Basil Liddell Hart, the originator and populariser of this phrase, is discussed later.

Europe, a cult of wishful thinking overtook policy makers, pundits and public alike. The slow and reluctant start made in exploring the opportunities for armoured warfare, together with all the debates associated with mechanising the army, and the equally slow process of adopting these lessons in the late 1930s, came about as a direct consequence of the absence of a perceived need. Government policy – the ten-year rule and the financial strictures which accompanied it – ruled out the requirement for preparations of any kind, insofar as the army was concerned – for a war anywhere but the empire. Wars in the empire, it was well known, were low-intensity affairs.

The fundamental problem was that if the army existed solely as an imperial gendarmerie, the more sophisticated requirements for doctrine on peer-on-peer warfare could at worst be ignored, at best treated as low priority. Yet between 1914 and 1945 the British Army was called on to fight two high-intensity wars. It had prepared as much as it believed it had to for 1914 with the publication of the Field Service Regulations, for example, but nevertheless struggled to create an effective army to deal with the challenges of this battlefield when it was required. In 1914, the BEF, though well trained and equipped, was just too small to make a significant difference. The changes necessary didn't come about until late 1916 at the earliest, after the mass casualties of the Somme demanded new thinking. Now, in 1919 with the experience of the war behind it, one would have hoped that retaining the knowledge that had been so expensively garnered and using this as the basis for the further development of military ideas and methodologies, would have been fully supported by the entire military and civil establishment, not to mention that of the various branches of government. But in 1919, with the threat from Germany now disappearing rapidly into history, the country and the army fell into a trance in which it ignored the possibility of there ever being a repeat performance to the war that had just passed. No consideration was made, therefore, for preserving and developing what had been learned.

The absence of doctrinal debate was dispiriting to many as early as the immediate post-war years. For Captain Eric ('Chink') Dorman Smith, adjutant of 1st Battalion Northumberland

Fusiliers garrisoned in Cologne as part of the British Army of the Rhine, the intellectual torpor of regimental life was disheartening. He sounded off in the battalion magazine, *St George's Gazette*, in 1923 (he could do so because, as adjutant, he was also the editor):

> The British nation is always engaged in detailed study of the lessons of the last war, as opposed to the more theoretical task of forecasting the methods of waging future wars. [For example, on exercise] our enemy never has tanks or low-flying aeroplanes [and] the query as to what does an attacking company do if it is attacked by a tank remains unanswered. The tactical scene of today puts the art of Jules Verne and Rider Haggard in the shade.[7]

A few years before, Major Bernard Montgomery was likewise dealing with his own frustrations as an early post-war student at the revived Staff College at Camberley. He hated the experience. To him and other veterans of the fighting the teaching seemed to be impractical nonsense unwedded to the realities of the modern battlefield. Why weren't they being taught how to marshal the armies of the future on the basis of what he, as the chief planner in 47th Infantry Division, had been doing in 1918? For Monty the primary task for the army was to be professional. War was a science; it required careful planning and management and thorough training. By reverting to being the imperial police force he thought the army was returning to amateurism, turning its back on the need to improve the seriousness of the profession. After his stint in Ireland Montgomery, by then 36, began marshalling his own ideas about how battle needed to be conducted in the next war. It wasn't revolutionary stuff, but he could never understand why it didn't seem second nature to everyone else.

The challenge for him was not to refight the Great War, or to seek fantastical solutions to the problems posed by modern, industrial warfare. Instead, it was with the lessons of that war and with the advantage of new technologies in artillery, tanks and aircraft, to work out how to fight the next one successfully. The debate about the nature of future warfare in the army during the 1920s and 1930s has often been reduced to one between the advocates of change and those – usually derided as cavalrymen – who were keen to hold on to their horses and embrace the past, rather than the future. Nevertheless, the idea that all army officers other than forward-thinking reformers such as Fuller and his acolytes were cavalrymen unwilling to give up the sword or lance, or crusty Colonel Blimps resolutely opposed to any form of military reform, is as inaccurate as it in unhelpful. Only a small proportion – between 6 and 7 per cent – of senior officers during this time were cavalrymen.[8] In fact, a considerable debate raged during the period, almost exclusively fought in military circles and on the periphery of the army's core business, about how the army should fight in the modern era. Much of this debate was taken up with the subjects both of mechanisation and of the tank, without much evidence that dissenting voices were ever hindered or suppressed. The public parts of the debate took place in the pages of the *Army Quarterly*, the journal of the Royal United Services Institute and the newspapers. Because of this it occasionally became acrimonious and 'political', with the radicals accusing those who did not agree with them in every respect as Neanderthals and worse.

The real problem was that to prepare the army for a future 'war of the greatest magnitude' against a peer opponent, clear direction was required from the Cabinet and War Office (because the army's direction, intent and purpose is ultimately a political decision) to determine what sort of army, in addition to that required for maintaining the security of the empire, was needed if at some time in the future a war broke out again in Europe. Part of this failure lay in the resolute refusal of most policy makers to consider that Britain might in due course be required to undertake a second continental commitment that century. In 1926 the CIGS, Sir George Milne, observed that the army could not even mobilise a single corps.[9] This wasn't necessarily a bad thing, he considered, as the army would never again be required to fight a European war. No one batted an eyelid at this statement, as its truth was self-evident. With that 1926 observation in mind, it is worth reflecting that at the time of writing the British Army cannot field even a single division, let alone a corps. Nor will it be able to do so in 2024 or at any point in the immediate future. And with what confidence can we say that we will not be required to fight a European war in the future?

Reverting to 1926, it is possible to distinguish three broad sides to the conversation about the future of the army in two inter-war decades. The first was that held by those who did not believe that the army needed to change much, outside of the normal improvements to weapons and technology that were required from time to time to enable it to carry out its task of protecting the empire from a second-rate enemy. They considered that the infantry, artillery and cavalry in its present form were sufficient for the army's needs, although the gradual adoption of the internal combustion engine in various parts of the army was thought to be helpful. This broad group could be described as 'conservatives', well-meaning men of good intent with considerable combat experience who nevertheless did not consider that they were on the cusp of a Revolution in Military Affairs. To this group, the tank accompanied the infantry, at the pace of marching troops, protecting them with the machine guns with which it was armed.

The second group were what we might term 'reformers'. This was a relatively small though influential group of men who recognised that 1918 represented a watershed in the concept of warfighting. They considered that while the infantry remained 'queen of the battlefield', infantry formations needed to be significantly enhanced by artillery, planes and tanks in order successfully to penetrate and exploit beyond defended enemy positions. They believed that firepower and the mechanics of mobility (e.g. tanks and mobile artillery) coalesced around the infantry, enabling the whole force to advance safely into the teeth of an enemy defence. These men tended to be pragmatic innovators, solving problems faced on the battlefield intelligently and imaginatively (as had been done so by Rawlinson's Fourth Army), but they rejected the argument that the future of warfare was based exclusively on tanks. They considered that engine-based mobility should be given to both the infantry and the cavalry, allowing commanders the flexibility that was provided by 'protected mobility'.

This group believed that the infantry (mounted in trucks or armoured vehicles) would remain the primary war-winning arm, though they accepted that it would fight closely integrated with armour, artillery and aircraft. In particular they considered that the infantryman of the future would be highly intelligent, superbly trained and equipped, able to use his initiative in battle in a way unknown to previous generations. He would no longer be a mere flesh and bone cog in a battle of 'mass'. The individual soldier needed to be a thinking soldier, not trained for the brutal slogging matches of the past but designed for intelligent and sophisticated operations of the kind seen to succeed in France in 1918. This type of soldier would be a technical specialist, requiring months of training. The 'new' army would likewise require well-trained, intelligent officers able to inspire their subordinates by the cogency and clarity of their plans, rather than expecting to lead through blind obedience. It would demand effective weapons and equipment (such as mortars, man-portable machine guns and radios) purposely designed and engineered for a new type of battle. Stealth, cunning and subtlety would be required in the execution of military operations. Reconnaissance, deception as to timings and locations of forces and elaborate planning would be essential. It would mean high levels of initiative of the kind that platoon commanders such as Duff Cooper had demonstrated in their assault on the Hindenburg Line. In the view of the reformers 'infantry' tanks remained tools that were required to travel at the pace of the infantry, while 'cruiser' tanks would go faster to undertake the traditional role of horsed cavalry, decoupled from and independent of the infantry. They considered that making the cavalry mobile with vehicles rather than horses didn't change their role per se, merely what they used to get from A to B. This broad group of intelligent practitioners included men such as Edmund Ironside, Henry Rawlinson, George Milne, Archibald Wavell, Bernard Montgomery, John Vereker (Lord Gort), Alan Brooke, John Dill, Claude Auchinleck, Harold Alexander, Charles Broad and Bill Slim.

The third group we can call the 'radicals'. This was a tiny number of noisy individuals, many of whom had served in the Tank Corps in the Great War, who believed they saw in 1918 an entirely new form of warfare, one that focused on the tank, almost to the exclusion of all else. These so-called 'apostles of mobility' such as Giffard Martel, Frederick Pile, J.F.C. Fuller, Percy Hobart and Basil Liddell Hart believed that armoured mobility could now give commanders a far more decisive advantage in war, making the infantry if not redundant, then superfluous or, at least, secondary.* They had seen the effect of shock action in 1918 first against the British Third Army at Amiens in March and then against Ludendorff's armies in the Hundred Days. They believed that the future of warfare would be that of massed tank armies wheeling behind fixed defences to defeat the enemy in the rear, destroying command, control and logistical infrastructure and thereby the enemy's ability to command its armies, rather than confronting its manoeuvre forces directly. Both Fuller and Liddell Hart came to believe that mobile formations of tanks were the key to future war. Fuller considered the triumph of 1918 to be the tank. Its utility was unarguable:

> Wireless, caterpillar tracks and armour – these are the three components of the new idea. You can talk to anyone, you can move over any form of open country, and you have cut the bullet out of the picture. Remarkable changes – yes; yet no more than the A B C of a new tactical language; for you have no front, your front may be anywhere; you have no reserves, your whole force is a potential reserve; you have no old-fashioned advanced guard, you have a moving armoured screen – the Light Tank Battalion, which, like the cloak of a matador, can be flicked from front to flank, from flank to rear, with incredible bullet-proof speed when compared to that of infantry.[10]

The view of this fraternity was that the tank – fast and heavily armed – could play as decisive a role on land as the bomber in the air, or the battleship at sea, an analogy Fuller used in a prize-winning

---

*See Chapter 10: The battle of the tank.

Royal United Service Institute military essay contest in 1919.[*] Fuller continued to believe that a large tank force striking deep into an enemy's rear to decapitate its command and control capability was the future of warfighting. In this latter formulation the tank was the modern horse, carrying out not just the traditional function of the cavalry but a strategic discombobulation role new to modern warfare. This, of course, remained an untested theory, in Britain at least. Fuller regarded the virtual destruction of the superb Tank Corps after the war as a moment of extraordinary self-harm. In 1918 a Mark V tank, for example, by being fitted with an aero engine, had achieved cross-country speeds of over 20 miles per hour. But, after 1919, rather than seeing this as a speed to which all accompanying arms and services had to follow, the army directed that the speed of tanks be reduced to that of marching infantry. It was, he expostulated, akin to ploughing 'a field with a tractor harnessed to a horse'.[11]

Fuller and Liddell Hart believed that a mental lethargy embraced the British military mind after the Great War. Fuller argued that the reversion to that of an imperial police force created a retrograde mental adjustment in the army, in which thinking about war, and about how to win at warfighting, became a pretence. If the Great War was an aberration in the history of mankind, and would never be revived, preparation for it need not be considered, or rehearsed. Exercises became, as Fuller asserted, 'mythological'. He pointed out that in 1932 exercises for the 1st Division in Aldershot, for example, were based on scenarios modelled on the Boer War.[12] Mental and intellectual preparation – thinking about war as a science – could be distracted by other more compelling likelihoods, such as imperial campaigning or military aid to the civil power, and the army's first duty, to prepare itself for the possibility of a 'war of the greatest magnitude' to defend the nation's interest, forgotten.

To what extent was he correct? Fuller's complaint was one that many others echoed, including Ironside and Montgomery. The problem was that of having officers capable of handling

[*]He had in fact borrowed this analogy from a paper written by Martel in 1915.

complex formations (that is, mixes of various units at divisional, corps and army level), with minds that had been formally trained and prepared for this purpose. Britain did not take seriously the intellectual training and preparation of its higher commanders during the inter-war period, with the result that the army faced a serious deficit of generalship in the early years of the Second World War. There was some validity in the argument that army officers weren't as inquisitive as they should have been during this period. Tuker, who lived through it, described it as 'The Great Ice Age' of military thought. Professor Brian Bond, while dismissing as exaggeration the claims of some that the army was unthinking, acknowledges that the army's innate conservatism was reactionary, 'in its deep-rooted reluctance to adapt to both technological developments and changes in the society in which it lived.'[13] Indeed, there is considerable evidence of the intellectually stultifying existence of most army officers in the 1920s–30s though there were a few – noted above – who were exceptions to the rule. The majority expressed little or no interest in military innovation, or what their friends and (potential) enemies were doing in the military sphere. It isn't surprising, as Bond points out, 'that British Army officers failed to appreciate the extent of the French Army's decline or the Wehrmacht's rapid development in the later 1930s.'[14] Field Marshal Philip Chetwode's much-quoted farewell address as Commander-in-Chief India in 1934 was perhaps the most devastating indictment of the military profession in this era by an insider, including the reflection that 'The longer I remain in the Service, the more wooden and the more regulation-bound do I find the British officer to be.'[15]

In the view of Major General John Kennedy, Director of Military Operations in 1939, at the start of the war 'the Army had no single soldier with war experience of high command.'[16] The problem was that Britain 'began with an Army which was trying – through no fault of its own – to expand too late, and with a nation which was rousing itself from a deep sleep as the lava began to flow.'[17] Kennedy's argument is correct insofar as it relates to the unique demands placed on the small peacetime army as it grew rapidly to meet the urgent new demands of war. But it doesn't fully exculpate the army from failing to prepare those commanders that it did have, for the potential challenge of war. As a result, the British generals who entered the Second World War in positions of high command did so without any direct or specific training for their roles. Those who succeeded often did so because they were self-taught. There was no systematic programme of training or instruction for any commanders at formation level (i.e. brigade and higher). Most senior commanders in the British Army, according to Sir Michael Howard, 'displayed in fact all the good and bad qualities of the Regular Army: excellent in looking after their men, brilliant in small-scale actions requiring flair, dash and leadership, but out of their depth ... and inept at higher levels of command'.[18]

Senior officers were expected by dint of character, on-the-job training and experience, bolstered for the few by attendance at the Staff College at Camberley, and even more rarely by the Imperial Defence College, to be able to master every problem thrown at them, without specific command training. It was not considered necessary to train officers to command anything larger than a platoon (the job of Sandhurst or Woolwich at the start of one's career) and any consideration of the 'art and science of war' was considered a Prussian eccentricity and not the proper fare for the British military tradition, which expected its commanders to develop their character riding to hounds, steeplechasing or on the sports field. This wasn't unreasonable: hunting and riding to hounds was regarded by many officers as being the perfect training for understanding and appreciating the nature of ground. But it patently wasn't the whole answer. No training in high command existed anywhere in the empire, a legacy of the fact that no formation doctrine existed for the deployment of armies in a 1918-type scenario. A 'Higher Commander's' course, run by General Dill in Aldershot in 1939, appears to have been the only attempt to meet this need. Fuller lamented that even large-scale military exercises ('manoeuvres') in Great Britain were abandoned as an efficiency measure in 1925.[19]

The result was that instead of doctrine dominating thought and behaviour, the personalities and private predilections of those placed in operational command became paramount. Command was defined on the basis of the personalities of the commanders and the various judgements and prejudices that formed their own individual military experience. This in fact had been the traditional British approach to command: the single military hero – the Wolfe, Wellington or Gordon, seemingly endowed with all the virtues of a superior race and schooling, securing the command of his army by virtue of the power of personal leadership. Personality is essential at lower levels of command, but personality alone is a poor preparation for dealing with the demands of high command in war, particularly when that experience is limited to peacetime soldiering or, at best, imperial policing.

Despite these criticisms, it is instructive to observe that the Staff College – for which 500 eager officers attempted selection every year – produced all the outstanding military commanders of the war, during both the years of defeat and the years of victory. Imagine the situation had the army enjoyed a more robust form of higher training for war for all commanders above sub-unit level (infantry company, tank squadron and artillery battery), closely wedded to a comprehensive set of ideas for the successful waging of high-intensity warfighting. It was not to be until 1988 that the then-Chief of the General Staff (CGS), Field Marshal Sir Nigel Bagnall, introduced the Higher Command and Staff Course for 30 selected potential senior commanders. The trickle-down value of this course has transformed the British Army of today, but complacency is always a danger.

In the circumstances of 1939, the failure of the first three years of war is understandable. Armies, like their leaders, take time to train, although this reality is hard to accept by politicians demanding immediate action and even quicker results. The problem, Kennedy observed, was that when war came Great Britain was 'still suffering from the effects of our national habit of neglecting to create an army until after the outbreak of war.'[20]

Britain's failure to select generals during peacetime on the basis of their military ability alone was clear to Brooke when he became CIGS in late 1941. 'Too many officers have been, and are being, promoted to high command because they are proficient in staff work, because they are good trainers, because they have agreeable personalities, or because they are clever talkers,' he complained. 'We must be more ruthless in the elimination of those who seem unlikely to prove themselves determined and inspiring leaders in the field. It is essential to select the best men to fill their places'.[21] His War Diaries are replete with complaints about the lack of suitable officers for command positions, which he credited to high losses in the First World War. On 8 October 1941 he wrote:

> It is lamentable how poor we are regarding Army and Corps Commanders. We ought to remove several, but heaven knows where we shall find anything much better. Post-war addition: This shortage of real leaders was a constant source of anxiety to me during the war. I came to the conclusion that it was due to the cream of the manhood having been lost in the First World War. It was the real leaders, in the shape of platoon, company and battalion commanders, who were killed off. These were the men we were short of now. I found this shortage of leaders of quality applied to all three fighting services, and later I was able to observe that the same failing prevailed among politicians and diplomats.[22]

Brooke's conclusions are unconvincing, however. The argument that 'the best' future commanders were all killed off in the First World War, leaving only the dunderheads to muddle through into the Second, is not logically sustainable. The real reason for the apparent ineptness of British commanders at the start of the Second World War is that they simply had not been trained or prepared adequately for high command during the years of peace.

Chapter 9

What is the army for, and how should it fight?

One of the difficulties the army faced during these years was that it had an inadequate system for managing innovation and change. Military doctrine needs a home, a place where the theory of warfighting can be translated into force structures, equipment, standard operating procedures and training. The army didn't have a directorate of doctrine or development to oversee this activity, which meant that where new thinking or the development of ideas did take place, it did so in an unstructured way, by zealots and often by outcasts. It was dependent on the predilections, good or bad, of its senior commanders, in field command or in the War Office. When these posts were occupied by men determined to continue the path towards improvement, consciously promoting and developing the creation of ideas across the army, the future of the army could be regarded as being in good hands. But there did not exist any form of formal or sanctioned laboratory where ideas could be tested or developed. In 1924 Captain Basil Liddell Hart suggested, in a paper calling for a New Model Army published in the *Army Quarterly*, that the army set up a tactical research department, which would work closely with a technical branch, to evaluate every type of military innovation. Nothing came of the idea, a lack of money being used as the excuse for inaction.

The great benefit of planning for future military eventualities is the opportunity it provides for war gaming various scenarios, as a means of calculating approaches, tactics and force structures. This degree of analysis also has the advantage of excluding those ideas, sometimes proffered by lobbyists, which are unrealistic or extreme. It is only in the absence of war gaming and tactical exercises that mad ideas are allowed to flourish. Equally, the experience of campaign planning would have allowed a fresh and open examination across all the arms and services of the art of the possible when it came to the effective deployment of men, weapons and equipment in battle. During the 1920s and 1930s a lack of money for large-scale training can partly be used to explain why new tactical concepts and ideas weren't rehearsed as they might have been.

The primary deficiency was in not having a single concept of operations in which the requirement for combined-arms manoeuvre across the battlefield was the foundation. In other words, if the doctrine had come first – working on the basis of what was known to have worked in 1918 – the types of equipment and weapons, and force structures (organisations), would have followed naturally. Not having a doctrine against which these new ideas could be tested meant that any innovations that were proposed by eager lobbyists had no means of being tested and evaluated and, if suitable, developed for service. If there had been such, the purpose of experimentation would have been clear. As it was, such experimentation as did take place had little far-reaching effect. One example of this was the advent of the Experimental Mechanized Force between 1927 and 1929 which appeared – wrongly –to be merely a sop to the tank enthusiasts (which included many infantrymen) who had been banging their drum since its invention and who needed a temporary sandpit to test their theories.*

Where could thinking about future war be undertaken? Without a dedicated unit designed for this purpose, there were two obvious

---

*The formation of the Experimental Mechanized Force (re-named the 'Armoured Force' in 1928) will be discussed in Chapter 10.

places. The first was the War Office, the second the Staff College at Camberley. The War Office had four branches: the General Staff, the Adjutant General's department, the Quartermaster's department and that of the Master General of the Ordnance. The General Staff, the brains of the army, had three directorates: Military Operations and Intelligence (DMO&I), Staff Duties (DSD) and Training (DMT). Ownership of pan-army doctrine could have been placed in any one of these departments. The problem was not one of organisation but of the appreciation by the army's leaders of the need for a unifying concept of operations in the first place.

In the early years following the end of the Great War, there was hope that the Staff College at Camberley would undertake this task. It did so for a short period of time only. Unfortunately, the Staff College was closed between 1914 and 1918, which meant that nowhere during the war was there any repository to capture doctrinal learning from the battlefield. When it was re-opened in 1919 the aim of the new commandant, Major General Hastings Anderson, was to examine and understand the lessons of the Great War. In the view of at least one student, Major Bernard Montgomery, as we have seen, he failed. Montgomery had wanted to understand what the British Army had learned from the Great War. He was not to discover the answer to this quite obvious question. The Staff College seemed entirely silent on the construction of a new 'future-focused' military doctrine built on what it had learned in France and elsewhere. When six years later he was invited back as a member of the Directing Staff, he jumped at the chance to put things right. He saw the opportunity to fill in the gap himself, even if no one else seemed to be doing it. The problem was that the traditional Staff College method, revived from 1919, was to study selected battles and campaigns from military history.[1] It was believed that history would enable the commanders of the future to apply the lessons of the past to the problems of the present, filtered, of course, through the Field Service Regulations and other special-to-arm pamphlets or training publications. Britain's army had never been built on a doctrinal framework (the Field Service Regulations were a post-Second Boer War afterthought); military history was its substitute.

To men like Montgomery, however, this approach was inadequate. Something much more radical was required, in which military doctrine was evaluated, constructed and taught.

This is not to suggest that the attempt was not made. For a short period of time, it was. General Edmund 'Tiny' Ironside took the reins at Camberley in May 1922 determined to add rigour and reality to the course. In fact, he wanted to convert the course into a 'War College' to explicitly instruct up-and-coming officers (the two-year course had 60 students in each intake) in how to fight and win the next war based on a scientific understanding of the wars of the recent past. To the teaching staff he invited one of the radicals: Colonel J.F.C. Fuller. This was something of a risk for Ironside, as Fuller was already regarded by many in the British Army as being a dangerous fanatic, obsessed with the idea that the tank would revolutionise warfare to the exclusion of all else. Both men tried to create a university atmosphere where combat-experienced officers could study the Great War and selected conflicts before it – such as the American Civil War – with a view to determining the lessons for the next. For example, Ironside used the intellectual stimulus provided by his appointment to write his account of the Battle of Tannenberg, the German defeat of the tsarist army in 1914. The experience was certainly appreciated by one of the outstanding students of the 1923 intake, Major Arthur Percival, later to be the commander of British troops in the doomed Malaya and Singapore.[2]

During Fuller and Ironside's tenures (the former left in 1925 and the latter in 1926) the Staff College was regarded as a beacon of new thinking about the art and science of war. However, in the years which followed something of a reaction set in, especially against Fuller, who had written his highly philosophical *The Foundations of the Science of War* while an instructor.[3] Fuller's theorising irritated the CIGS at the time, Field Marshal Lord Cavan, who refused him permission to publish his written work. For this, Cavan has been castigated as a moribund reactionary unwilling to engage with new ideas. The book was published only when Cavan had retired in February 1926. But Cavan had a point. On examination the book is highly philosophical. Fuller was attempting to alchemise warfare, distilling its core principles into base ingredients. This is commendable for a military scientist. But Fuller was attempting to use this to push his tank barrow, and it was this – together with his over-theorising about warfare – which irritated the War Office. Opposition to Fuller didn't occur merely because he was a radical with respect to the future of the tank. Field Marshal Archibald Montgomery-Massingberd, CIGS from 1933 to 1936, also had little time for Fuller's prognostications. Montgomery-Massingberd had an impressive though mixed record as an innovator, having been Rawlinson's deputy in Fourth Army during the battles of 1918. He was a leading advocate of the mechanisation of the British Army in the 1930s. But he disliked Fuller in part because he believed military affairs to be far less theoretical than Fuller made them, and thought that Fuller was attempting to sell short cuts to success. His failure, along with most of the senior officers of that era, was to lack the imagination to recognise that future war – even as far away as ten or twenty years hence – would require preparatory action *now*. His view about the possibility of anything such happening while he was in uniform can be summed up in a speech he gave in February 1933, soon after taking the helm from Milne: 'I will venture to say that the Army is not likely to be used for a big war in Europe for many years to come.'⁴ Like most other such predictions during this period, as we now know, this was a classic case of the ill-founded wishful thinking that bedevilled the era. The greatest curse of peacetime military leaders is for them to assume that they exist only to manage the present, rather than also preparing for the uncertainties of the future.

A further complication in an army without any formal doctrinal bedrock beyond the Field Service Regulations is that few in senior positions understood what doctrine was. Even the two most prominent theorists of the age – Liddell Hart and Fuller – spent a lot of their time discussing and refining the principles of war. These can easily become academic and theoretical and need translating into actionable templates for military decision-making by soldiers at every level of command, from infantry section upwards. The Field Service Regulations were too remote from practical soldiering to enable a divisional commander, for example, to understand the most effective way to manage an opposed advance to a military objective. When CIGS Lord Cavan told the Cabinet in January 1924, 'we require no plans of campaign (except for small wars incidental to our imperial position)', he did so because he believed that the British Army's traditional adaptability would solve any problems when they arose.[5] Although his specific point related to planning for possible war, he was reflecting a common misapprehension among senior officers at the time that every military problem could be solved by recourse to the intelligent pragmatism of military leaders. General Sir William Robertson (CIGS 1915–18) put it like this in a newspaper article in 1925:

> [W]ars have invariably been accompanied with many unexpected developments and the conqueror has owed his success more to rapidity of mobilisation and adaptability to circumstance, i.e. to common sense, than to any prevision on his part, either as to the character of the particular war in which he was engaged, or the effects of the weapons employed in it.

When, in 1931 Sir Archibald Montgomery-Massingberd was questioned about a detailed plan of operations for a prospective war in Europe, he observed that it would be a nugatory exercise as it was impossible to foresee future developments. In other words, planning for the conduct of any future military commitment to Europe could safely wait until such an eventuality materialised. There was no requirement to plan for it in advance. As Professor John Gooch has observed this 'was not the stance taken by European general staffs of the day.'[6] An alternative approach, one dictated by the requirements of doctrine, would have been to ask a set of questions, such as:

1. What is the nature of the threat we are likely to face in Europe?

The attack on the St Quentin Canal in September 1918 was one of the great battles of the British Army in the Hundred Days campaign. (Getty Images)

Tanks in their 'stables' following the Battle of Hamel, July 1918. Their performance at Bullecourt in 1917 had caused scepticism among Australian soldiers, but these new models were faster and more manoeuvrable. (Getty Images)

The new type of warfighting in 1918 encapsulated in a photograph of the capture of Grevillers on 25 August 1918. (Getty Images)

The art of manoeuvre was sufficiently well developed during the Hundred Days campaign to include supply tanks advancing with forward troops. (Getty Images)

Their long war finally over, happy members of the Royal Field Artillery arrive in Dover following their return from Salonika, Greece in January 1919. (Getty Images)

Soldiers of the London Scottish Regiment march down Fleet Street during the Victory Parade. After such a long and terrible war it is unsurprising that few could stomach the thought that the world would ever contemplate another. (Getty Images)

The Great War was an imperial affair. Here, Gurkhas of the Indian Amy Contingent march through Admiralty Arch to the Cenotaph during the Victory Parade. (Getty Images)

Tanks, a potent symbol of British military power in Germany, being inspected at Düsseldorf by General Thomas Moreland, commander-in-chief of the British Army of the Rhine. (Getty Images)

British troops with an ad hoc form of vehicular mobility, equipped with a Lewis gun, on guard at the Jaffa Gate, Jerusalem, 1920. (Getty Images)

From the Great War abroad to civil war at home. Two members of Sinn Féin are arrested following a raid on the Ministry of Labour offices in Dublin during the Anglo-Irish War. (Getty Images)

A detachment of British Lancers crossing the Rhine bridge into Cologne, Germany, part of the Allied occupation of the Rhineland to enforce the Treaty of Versailles. (Getty Images)

British troops leaving southern Ireland at the conclusion of the Anglo-Irish Treaty, signed in December 1921. (Getty Images)

In a spill over from the Great War, British troops found themselves in Gallipoli, guarding the road to Istanbul, during an attempt by Kemal Pasha to overturn the Great War settlement. (Getty Images)

Imperialism on the cheap. The RAF persuaded the government that recalcitrant tribesmen could be kept in their place by the use of aircraft dropping aerial bombs. (Getty Images)

In an enduring military commitment to Egypt, an armoured car patrol makes its way along the Nile at the Old Cairo Quay. (Getty Images)

Recruitment was a problem in the 1920s. This postcard is from the Army Recruiting Stand at the British Empire Exhibition at Wembley in 1925. (Getty Images)

In a classic Military Aid to the Civil Authorities operation, British troops deploy in London to protect convoys during the general strike of 1926. (Getty Images)

The British Army stationed 55,000 troops in India between the wars to operate alongside the Indian Army in protecting the North West Frontier and offer Military Aid to the Civil Power. (Getty Images)

Men of the Scots Guards leave Southampton en route for service in China following the threat by Cantonese forces to British interests in the International Settlement at Shanghai. (Getty Images)

A Royal Tank Corps Armoured Car Section patrols an area of the International Settlement in Shanghai. (Getty Images)

A caricature of the British Army between the wars, neither one thing nor the other, uneasily managing the cusp between the old and new. (Getty Images)

A posed photograph of men of the Royal Tank Corps on exercise. The development of the tank was delayed by confusion over the best way of deploying them in battle and by a lack of funds. (Getty Images)

British troops return to the UK following the end of the post-war commitment to garrison the Rhineland. (Getty Images)

A British soldier in a Vickers Carden Loyd Machine Gun Carrier Mark VI tankette on a training exercise at Pirbright, Surrey in August 1929. (Getty Images)

Loyd carriers and horses on exercise together at Pirbright in August 1929, representing the nexus between old and new in warfare. (Getty Images)

Men of the Royal Fusiliers march through Wiesbaden, the last of the British Army to leave the Rhine. (Getty Images)

British troops at the Damascus Gate, Jerusalem, during riots in Mandated Palestine in August 1930. The commitment to Palestine was universally detested. (Getty Images)

British and Indian troops deployed on Military Aid to the Civil Power tasks in India dispersing rioters in Peshawar. (Getty Images)

Officers of the Lincolnshire Regiment, part of the BEF, somewhere in France in early 1940, taking part in a divisional exercise during the Phoney War. (Getty Images)

Men of the BEF in a trench in March 1940, in a picture that could easily have been taken, except for the presence of the Bren gun, during the Great War. (Getty Images)

British and French troops waiting on the dunes at Dunkirk to be picked up during the Dunkirk evacuation, 26 May–4 June 1940. (Getty Images)

Troops evacuated from Dunkirk in Royal Navy destroyers berthing at
Dover on 31 May 1940. (Getty Images)

A German photograph of one of the roads leading to Dunkirk after the final evacuation
of the BEF in early June 1940. (Getty Images)

2. Will this require us to conform with the strategies and approaches of a partner, such as France? If so, what will be the tasks we need to place on our deployed expeditionary force?
3. How will the military threat evidence itself in practical terms?
4. What is the best way of facing that military threat, either alone or with our partners?
5. How will this determine our fighting strategies, organisational plans and the physical deployment of our forces?
6. What do we need to do to sustain these forces if the war lasts more than three or six months?

Clearly, a thorough evaluation – as yet not undertaken – of the battles between 1916 and 1918 would have allowed the War Office to do this. The absence of any doctrine to underpin operations in 1940 was a significant factor in their failure. One further example will suffice. The absence of doctrine with regard to the defensive deployment of a modern army built on the experience of 1918 meant that when the second BEF was deployed to France in 1939 to align itself on the northern edge of the French Army, no serious conversation took place with the French Army about the need for an *area* defence of northern France, rather than a *linear* one. An area defence gave up the idea of heavily defended lines and spoke instead of defended bastions – interlocking by fire where possible, strung out deeply into the defended territory to absorb and separate an attacker. The French plans for a linear defence system similar to that forced on the Allies at the end of 1914 was accepted without much demur.

The 280-mile-long Maginot Line in France, construction of which began in 1929, was a classic example of the folly of such a concept at the operational level of war, but nowhere in the Field Service Regulations, or anywhere else for that matter, is the fundamental difference between a linear defence and an area defence laid out. If it had been, some of the errors made in 1939 and 1940 in conjunction with the French defence of northern France and Belgium could have been avoided. The problem with a linear model is the reality that behind the (presumably) thick outer crust of the front line of defences are additional rear or support lines of differing strengths. The bastion model, in the concept of defence in depth, is that each of the interlocking bastions is equally strong.* The Indian Army's Lieutenant General Sir Francis Tuker railed against the linear mentality in his *The Pattern of War* (1948), blaming the generation of military intellectuals after 1918 for refusing to ditch the concept of the line of static defences as being outmoded and dangerous. With the Maginot Line as the epitome of this (false) conception relative to a mobile army (an 'army of manoeuvre' as he calls it) he describes the alternative available to the Anglo-French armies in 1940 – defending many specific areas across a wide geography, a subject we will revisit in Chapter 14:

> The most economical form of [strategic level] defence is a defence of areas ... By holding areas one need place but a nucleus garrison in those quarters against which an enemy's attack is least expected and can fill up behind the crust of those quarters where it has been finally determined that the enemy will make his main thrust. That is to say, since the Allies had no great army of manoeuvre with which to oppose their enemy, they should have faced him on the frontier with their mobile outposts instead of with a complete static linear defence. Behind these mobile outposts should have been the army of manoeuvre and it should have been based on well-organised and well-stocked areas of fortress defence. Such areas would be about Calais and Dunkirk, Lille, St. Quentin, Abbeville, about Paris, Rouen, Le Havre,

---

*It should be noted that the inter-war theory of defence, which reasserted a pre-1918 approach to defence in which lines are pre-eminent, was fundamentally the NATO approach to defence in North West Europe until the reforms initiated by Field Marshal Nigel Bagnall in the 1980s. Both authors lived through and experienced these reforms at the time in the deployments and dispositions of the British Army of the Rhine (see Chapter 17).

Caen, about Lyons and so on. Classic war demands the defence of fortresses by the *levée en masse* of its citizens, thus releasing the army of manoeuvre in full for its task ... It is possible that two great fortress-areas strongly held near the frontier, one in the north of France, the other about Metz, Strasbourg, Nancy, would have given the army of manoeuvre a sufficient power to prevent by threat, or by ultimate action against the enemy's flank and rear as he passed through, any considerable invasion of French territory.[7]

A study of the written output of the British Army in terms of instructions and pamphlets demonstrates during the inter-war period that at one level the army wasn't asleep. It was not as though doctrine weren't being written. The Field Service Regulations arrived in periodic revisions, in 1920, 1923, 1924, 1929, 1930 and 1935 and single-arm manuals were also produced. The latter articulated organisational principles or battle procedure. The Field Service Regulations comprised approaches to warfare in terms of principles, rather than (always) specific detail. On the whole they were sensible articulations of military art and science. But there was a serious problem with them. Understanding how well an army appreciates its doctrine comes not just from measuring the number of pamphlets it produces, but in assessing the degree to which the army lives and breathes by this teaching, and the extent to which force structures, equipment, training, preparation and rehearsals ('exercises', field, sand table and table top) follow. Doctrine ('what the army teaches itself about how to fight war'), force structures ('how the army structures itself to fight, in brigades, divisions, corps and armies'), the equipment programme ('what the army acquires to help it to fight, in terms of weaponry, vehicles and other technology') and campaign plans for various anticipated contingencies operate together as part of an integrated system. The system is based on a set of fundamental principles which together encapsulate an approach to warfighting.

For example, the modern British Army's approach to battle is called the Manoeuvrist Approach and is set out in the seminal document, 'British Military Doctrine' published in 1988; it is the foundation of the teachings of the Higher Command and Staff Course which, since the course's inception, has now influenced the thinking and the development of the whole army.[8] During the inter-war period the lack of such a system prevented the achievement of doctrinal coherence, with a variety of challenges remaining unresolved.

Despite the excellence of the 1935 Field Service Regulations, for example, which described management of war at the higher or grand strategic level, there remained three primary deficiencies in the articulation of British military doctrine. First, there was no overriding imperative to prepare, train for and fight a high-intensity war in Europe, nor any overall conception of how this type of battle needed to be fought. As the possibility of war in Europe increased, the army remained committed, ideologically and structurally, to the defence of empire. This lack meant no single unifying method of fighting that would allow commanders to plan, organise and deploy for each phase of battle, using a set of well-known and tested tactical methods that integrated all arms and services engaged in the fight. The need for uniformity was in fact well understood and commented on regularly at staff conferences. But this missed a fundamental point. Uniformity of tactics and principles is only part of the story: it requires a uniformity of operational concept. The author of Field Service Regulations 1923 grasped this perfectly, explaining that the purpose of an expeditionary force was to defeat the enemy 'by sustained and vigorous offensive action, which implies a war of movement.'[9] Indeed, the army would have been well advised to make this its *raison d'être*. Instead, the 1923 Field Service Regulations admitted that it was impossible to retain an army at home equipped and organised 'to meet the requirements of a great war' and asserted that the only alternative in the circumstances of peacetime was to have an army 'suited to average rather than exceptional conditions ... capable of modification to suit the special requirements of any particular campaign, and of rapid expansion to meet the case of a grave emergency.'[10]

These, as it turned out, were statements of policy, not of doctrine. Creating a doctrine to reflect the needs of high-intensity warfighting would have been relatively easy to undertake. A place to start could have been to develop the excellent *The Division in the Attack – 1918*, expanded upwards to army level, and outwards to incorporate the role of the division in defence as well. Another would have been to work backwards from the operational victories of late 1918 to ask the question: 'How were these achieved?' Setting out, in chapter and verse, the methods and processes that led to battlefield success would have served to answer the question. These answers would need to be laid out and codified for every branch of the service so that all ranks, from top to bottom, could then be trained in this specific approach to warfare. Without a structured explanation of how battle was to be managed, by each arm and service, working alongside each other in an integrated manner, the likelihood is that training will be misdirected, and open to the whims and inclinations of each individual commander. Uniformity of operational approach was clearly the most fundamental requirement, enabling army, corps, divisional and brigade commanders to follow the same approach to planning and delivering successful combat operations. This uniformity is reinforced by means of regular training and repeated rehearsals, in the sandbox and in the field.

Secondly, despite the attempt in 1923 to create a set of Field Service Regulations designed to build a future BEF there was no operational doctrine for campaigning. The 1923 Field Service Regulations did an excellent job of laying out the administrative principles for launching an expeditionary force but did little to explain how the subsequent campaign should be commanded, organised or fought. There was clearly a place for a revised version of *The Division in the Attack – 1918*, perhaps labelled 'The BEF in Attack and Defence' in the context of an operational ('campaign') plan.

Equally, just because a subject is explained adequately or well in a training pamphlet does not mean that it is understood widely, or taught extensively or well. Many pamphlets were the product of the efforts of well-intentioned single-arm lobbyists, determined to improve the quality of that arm's skills, equipment or operational purpose. Such were the Infantry Training Manuals or Lieutenant Colonel Charles Broad's excellent 'Purple Primers' of 1929 and 1931 (see Chapter 10), setting out ideas for the structure of armoured organisations following his experiments with the Tank Brigade. Although helpful, these conversations tended to be discussions within an echo-chamber, not engaging with or developing doctrine across the army.

However, the primary problem of the inter-war years wasn't the inquisitiveness or otherwise of its officers, but the political determination of successive governments that Britain's national interest did not lie in Europe, but exclusively in its empire. It was a strategic assumption based not on geopolitical realities but on popular revulsion at the scale of the harm done to the country by the Great War. The army could defend the empire as it had traditionally done, with the Royal Navy protecting the vast commercial sinews that held its sea lanes of communication together. Meanwhile, if Britain were threatened by a continental neighbour, it could be deterred and Britain if necessary defended by the new bomber fleets which a wide range of theorists were propounding would transform the nature of warfare.

It is important to criticise the army for failing to think carefully enough about the future of warfare, as we have done. But it is equally important to criticise the government for its egregious failure to acknowledge that were the balloon to go up in Europe a modern army would be critical to enabling the country to respond adequately to this challenge. Instead, it was deliberate government policy by successive administrations after 1919 to downgrade the army to a point where in a short matter of years the great fighting machine created in 1918 had become as distant to the reality of modern warfare as the Peloponnesian Wars. The government embraced the post-1918 peace dividend with no thought to the consequences of denuding the country of its fighting strength. It persuaded itself that the like of the war that had just ended would never occur again. It was, on reflection, an extraordinary abrogation of the government's responsibility when it determined that British national interest at home and in Europe would continue to be protected by the Royal Navy, increasingly by the Royal Air Force, and only peripherally by the army. Indeed, during this period the air defence of Great Britain was considered to be a more important task for the army than the need to prepare for warfighting in Europe, Brian Bond noting that 'anti-aircraft guns and air defence generally received a much higher priority than tanks.'[11]

The government failed to prepare for the fact that the British Army might once again have a requirement to support France in the event of a breakdown in European security. Ruling out such a commitment until it became inevitable as the result of German military aggrandisement demonstrated the foolhardiness of this strategy. Britain and its army needed to be closely linked with their allies in Europe so as to ensure that the peace of 1918 was preserved. A commitment to sending a modern, integrated warfighting force to Europe was a *sine qua non* of effective European collective security as was largely the case again at the end of the Cold War. With the withdrawal of the Rhine army in 1930 Britain, on the basis of a promise from Germany that the Rhineland would not be remilitarised, no longer had the forces able to enforce the new status quo. The Rhineland commitment for Britain had run its natural course, but the drawdown of the garrison was taken as a saving, reducing yet further the size of the British Army. Its handmaiden – failing to create an army able to fight as it had done successfully in 1918 – made Britain fundamentally weaker and incapable either of standing up to the bullies in the European playground or of fighting them when necessary. Indeed, had Britain had an army structured for peer-on-peer warfighting, this might well have served as a deterrent the British government could have used in later years to counter Hitler. If the deployable forces had existed, and if the British government had had the political determination to deploy these forces to counter the German reoccupation of the Rhineland in 1936, followed by the territorial aggrandisement of Austria and Czechoslovakia, it is very possible that Hitler would have been forced to back down, and reconsider his options. It is plausible that, had Britain and France entered into a military alliance before this (as France had requested) and responded with even a military operation against Germany, it might have worked, and Hitler might have been overthrown by his enemies in the General Staff.[12] The principle of deterrence is that one has a sufficiently capable military force to prevent an opponent launching military action, with a government willing and able to use this force in the protection of its national interest. In the Rhineland, Britain and France lacked the foresight or political imagination to accept that their respective national interest was imperilled and so did nothing to counter Hitler. In Britain's case there was no appetite for military action – many sympathising with Germany – but likewise the British Army was undoubtedly incapable of confronting or repelling the Germans even had it been deployed. It is no surprise that the lack of a commitment to European security had a direct and fatal impact on the development of a warfighting doctrine in the British Army, together with the equipment and weaponry necessary to deliver it. It is likewise no surprise that Britain and the British public failed to understand the implications for the balance of power in Europe if it refused to stand by the agreements already in place that established and enforced the security status quo over the Continent.

A major problem of the inter-war years in the senior echelons of the British Army was therefore one of *imagination*. Who in 1919 would have imagined that Germany would be overwhelmed by a destructive political and cultural virus that would allow it to behave as it did from 1936, and invade Poland in 1939, and Norway and the Low Countries and France again in 1940? Such a suggestion would have been entirely preposterous. Anyone suggesting such foolishness during the 1920s would have been consigned to the lunatic fringes of the political spectrum. Yet the unimaginable happened. It was a core requirement of the Chiefs of Staff and the War Office to *imagine* what might be, and prepare for it, rather than to merely hope for the best. The failure of Britain, both its government and its army, was to think that the unimaginable was not possible and thus to be critically incapable of responding when it did happen. In practical terms it meant that the British Army was not structured, organised or equipped to fight a peer adversary again in a major war. Just because a continental enemy in the 1920s was unidentifiable should not have meant that no such enemy would not one day emerge. It is the duty of the government, then as now, if they take the security of the nation's interest seriously, actively to imagine the unimaginable. In January 2022, a war in Europe was both unthinkable and unimaginable, but on 24 February that year it became a reality. History has an uncomfortable habit of repeating itself.

Chapter 10

The battle of the tank

For much of the 1920s and 1930s the primary doctrinal debate in British military circles centred on the nature of mechanisation, the extent of its adoption and its implication (if any) for land forces as a function of mobility on the battlefield. At one level mechanisation entailed merely the replacement of the horse with the internal combustion engine. For most people, however, the focus of the debate extended beyond this to one which we might call protected mobility – that is, traversing the battlefield without incurring the sort of casualties to unprotected infantry suffered in the Great War. A minority, as we have seen, saw mechanisation as a step beyond even this, considering it to be a means of presenting the tank as *the* means of avoiding the hard infantry fighting that had characterised the last war altogether. In this new world of their imagining tanks would sweep trenches into the dustbin of history by attacking not the forward enemy troops but the command, control and logistics apparatus in the rear.

The discussion almost didn't happen. Immediately following the end of the war, the Tank Corps was very nearly removed entirely from the British Army's order of battle, like the Machine Gun Corps which had been disbanded in 1922. It was saved, though in a very much reduced form of four battalions, with an additional 12 (soon reduced to ten) armoured car squadrons scattered across the empire. Despite these tiny numbers, the inter-war years saw the initially promising development of an effective tank, accompanied by valuable experimental work examining formation-level (brigade and division) structures and tactics in 1927 and 1928. Additionally, some organisational doctrine came along in the form of Lieutenant Colonel Charles Broad's first 'Purple Primer' in 1929. His second 'primer' in 1931 pointed directly to the principles of armoured formation manoeuvre – or warfighting – which was taken up by Field Service Regulations Part III in 1933, an analysis of the likely shape of future combat between armoured forces. Unfortunately, by the late 1920s tank development effectively stalled and the doctrinal field was fought over by those adherents of Fuller's argument, which saw tanks operating on their own as a form of modern, mechanical horse, and those who still saw the tank to be essentially an infantry support vehicle.

A common view today of the tank during the inter-war period is that it was defended against its detractors by brave modernisers (the 'radicals' of Chapter 8) who alone and in the face of significant prejudice were determined to drag the British Army kicking and screaming into the 20th century. The evidence, however, suggests that something close to the opposite occurred. There seems little doubt that the ideological battles that took place over the role and purpose of the tank in fact hindered both its technical and doctrinal development. The dominance in the debate of lobbyists like Fuller and Liddell Hart helped to create a groundswell against unproven theories and excited prognostications. In fact, the British Army never stopped *believing* that tanks would play an important role in the next war. Contrary to Liddell Hart's 1965 memoirs, the records shows that the War Office and successive CIGS encouraged and supported the development of the inclusion of tanks in the British Army's order of battle. The problem was always the lack of money for the tank's development and the tiny numbers involved, as well as – for most of this period – an absence of any army-wide agreement as to how these vehicles would be deployed in battle. What role would tanks play in warfighting? How would they inter-operate, if at all, with other arms? To what extent were other arms required to be mechanised as well? These questions were never satisfactorily answered before war came again in 1939.

In 1918, with no end of the war in sight, Fuller had authored a plan (borrowed from an earlier paper by Major Giffard Martel) to deploy a massive force of 4,992 tanks to smash through the enemy lines and to operate similarly to a naval fleet in the enemy's rear, attacking the enemy's command system. This 'Plan 1919' was obviated by the end of the war, but he captured these ideas in a book called *The Reformation of War* in 1923.[1] He suggested that an armoured advance would be accompanied by a fleet of 500 bombers simultaneously sweeping through the air to cause panic and mayhem among the enemy population. Both aircraft and submarines would also debouch small tanks into enemy territory. To most sensible soldiers these ideas smacked of fantasy, grounded not in the proven combined-arms effect of 1918 but in the unproven expectations of a second Cambrai of 1917. In the 1920s Fuller was given considerable opportunity to present his ideas to the army. The CIGS at the time – Sir George Milne – made him his military assistant in 1926 (after Fuller had left the Staff College) with the specific task of setting up the army's Experimental Mechanized Force in 1927, although the relationship between the two men did not prosper.

In part, the Experimental Mechanized Force arose from a concern of Milne's that if a resurgent Germany forced another continental commitment on the British Army a second BEF, using troops existing in Britain, would be required. Those troops did not exist, however, outside of the Cardwell 'home' units, most of which were in recruitment and training mode to support the commitment to imperial defence. When, in 1927 a divisional-strength expeditionary force was required for Shanghai, it proved difficult to find and difficult to organise. In the end troops were pulled from other duties, embarked and despatched to China without any preparation for war or any form of formation training. Likewise, it was an entirely infantry force.[*]

---

[*] Twelve infantry battalions, two of them Indian, in three brigades (13th and 14th Brigades and 20th Indian Infantry Brigade).

But without a clear doctrinal framework, formation exercises to test the limits of mechanisation proved aimless. An exercise was undertaken on Salisbury Plain in September 1925 to test the issue of mobility between mixed infantry, tank and cavalry units. It proved to be something of a fiasco but was one of the factors that led to the creation of the Experimental Mechanized Force in May 1927. Milne offered Fuller the opportunity to take command of the Experimental Mechanized Force but Fuller refused, on the basis that he would also be responsible for command of the Tidworth garrison, an administrative job he believed clashed with the need to get the most out of the Experimental Mechanized Force trial. In retrospect it was a foolish decision, as it meant that the one person who had been advocating more than anyone else the need to establish mechanised combat forces in the British Army was to play no role in directly influencing the result.

The Experimental Mechanized Force was to comprise half of the tiny Royal Tank Corps* (i.e. two battalions of the four in existence at the time), an infantry battalion, an 18-pounder field artillery regiment together with a 3.7-inch mountain gun battery, and an engineer field company. The concept was to create a fully integrated force of mobile ('mechanised') troops with a balance of tanks, machine-gun carriers, infantry and artillery. Unfortunately, it did nothing of the kind. Managed more effectively, the Experimental Mechanized Force could have been the test bed for the transformation of the entire army. For this to happen, however, would have needed at least three things to occur. The first was the funding to experiment as widely as possible with a range of tracked and wheeled armoured vehicles for all arms – tanks, infantry, armour and combat support, artillery and engineers, and combat service support, logistics – in order to determine the best vehicles, mixes of equipment, types of weapons and tactics. The second was agreement as to the extent of combined-arms integration across the army. Was this new combined-arms grouping, and mechanisation,

---

*It became the Royal Tank Regiment in 1939.

to be for the whole army, or merely part of it? The third, even more important perhaps, was a warfighting doctrine to ensure that these new formations could operate successfully in battle.

In the Southern Command exercise in 1927 the Experimental Mechanized Force demonstrated that it could operate successfully against an infantry division three times its size. But an unusual form of self-emasculation seemed to embrace the War Office over these efforts. The CIGS on 8 September 1927 told the men of the Experimental Mechanized Force that a complete change of mental outlook was required in preparing for a future war, by which it was presumed he meant a war against a continental enemy. The title of Milne's speech was 'For if the trumpet gives an uncertain sound. Who shall prepare himself for the battle?'[2] In it he urged the army to think differently and to embrace the reality of mechanisation, armoured warfare and a rethinking of traditional infantry-based tactics. Yet avoiding these realities is precisely what he and his successors had been doing and continued to do. He prefaced his remarks by pointing out that financial stringencies meant that what the army was describing as 'mechanisation' was going to be a long, slow process. He warned those who saw the successful result of the exercise that it was merely experimental and that in any case future development was constrained by the budgets available. Most tellingly of all, he added that the programme of mechanisation was going to be a slow one, 'so as not to upset the traditions, the *esprit de corps*, and the feeling of the Army as a whole'.[3] The phrase 'mechanisation' was itself a problem, for which Milne was unable to provide an unequivocal definition. Did it mean the development of an operational combat force of armoured vehicles, or merely the substitution in the army of the horse for the internal combustion engine? No one was entirely clear. History has shown that it was the latter.

The problem, demonstrated by his comments, and despite his support for the Experimental Mechanized Force, was that Milne wasn't clear himself what he wanted the new force to demonstrate, except that it was expected to have a minimal impact on the Cardwell-shaped army. He wasn't a Henry Wilson, knowing exactly what he wanted and working all out to secure it. He appeared overwhelmed by the magnitude of the intellectual and organisational task and the financial constraints upon him. The confusion was evident to the soldiers on the ground. The Regimental History of the 3rd Carabiniers, one of the horsed regiments participating, tells of exercises run through with uncertainty of purpose or strategic direction:

> In 1929, on Salisbury Plain, the Cavalry Brigade was used for the most part in opposition to the Experimental Armoured Force, to bring out the merits and limitations of both. To launch cavalry against armour seems a strange way of determining their relative merits and demerits and the result, as may be imagined, merely caused bewilderment. On the whole the cavalry seems to have had the better of the encounter for the armoured force was broken up. In the cavalry, 'the great problem to be solved is the harmonisation of horse and motor in the matter of speed, cross-country performance, supply, etc'. It was a problem which never was solved. Horse and motor have such divergent characteristics that to attempt to 'harmonise' them was pure folly.[4]

Then, at the annual Staff Conference in January 1928, a New Model division was proposed, designed to fight a major war in which the separate tank brigades were to be broken up and a tank battalion and armoured car company added to each infantry division. This new division would be the first truly combined-arms division, a revolutionary suggestion that would have had radical implications for the way in which the army structured itself. But it was not to be. The New Model division fell at the first hurdle. A lack of money was given as the reason: there was not enough in the kitty to allow the army to press ahead with the development of two types of division, an infantry one and a combined infantry/tank version. Such modernisation as was agreed would take place at an evolutionary rather than a revolutionary pace. Any changes would have to take place without upsetting the Cardwell settlement, as that, it was believed, would prejudice the defence of India. In any case the Treasury presented the argument that it was premature to start production of armoured vehicles when the army wasn't yet certain what it wanted.

The argument that doctrine would develop alongside experimentation was not robustly presented by Milne, and he gave in to pressure. As Jackson and Bramall have wryly observed, it 'would have taken a Marlborough or Kitchener and an obvious threat to the country's security to prize open the Treasury grip on military expenditure in the mid-1920s'.[5] Fuller wrote disgustedly that instead of building up a 'truly mobile little army ... we would not do so'.[6] Accordingly, when in 1931 the old-style 1st Infantry Division mobilised for exercise, it did so almost exactly as it had done in 1899, 'with 5,500 horses, 740 horse-drawn vehicles and the old horse-ambulances of the Boer War period'.[7]

So, while the idea of the Experimental Mechanized Force was positive, its execution left much to be desired. The army learned much from the experience, but it suffered from a lack of urgency, and the assumption that tanks were an addition to the old style of warfighting rather than an opportunity to create something quite new. The policy as Milne presented it was to make haste slowly, recognising the financial realities of Britain's economy, a point that the 'radicals' were accused of never fully appreciating. The radicals still had a point, nevertheless: change was needed. The army was crying out for a combined-arms division. With plenty of evidence to support their view in terms of what they had seen of the Experimental Mechanized Force they feared that the War Office did not understand how to sponsor change while simultaneously attempting to manage existing operational commitments. The radicals, of course, by promoting too much change too quickly, undoubtedly overwhelmed the capacity (and often the good humour and equanimity) of those in command of both policy and purse strings. A middle way was required, combining a recognition in the War Office for the need to change and a methodology for achieving it.

The problem was that, in the urge to mechanise, the ultimate purpose, apart from going faster than horse or foot power allowed, was absent. The ultimate purpose of mechanisation should have been to allow the army to have flexibility to deliver operational manoeuvre. Vehicular mobility provided the opportunity for commanders to manoeuvre their forces against the enemy, at both the operational (i.e. as part of a wider or campaign plan) level of war and at the tactical (i.e. directly on the battlefield). Though mechanisation was clearly important to achieve this, it wasn't an end in itself. Its purpose was – or should have been – framed by operational doctrine. In all the debates of this period this was the point that was not always fully understood. If one started with the doctrine of operational manoeuvre, mechanisation would follow.

The difficulty for the War Office was that the so-called 'apostles of mobility' – Liddell Hart and J.F.C. Fuller in the main – preached a doctrine of extreme manoeuvre that would never be accepted by the War Office or the Treasury, the one for the extremity of its ideas and the other for its cost. The view of the 'conservatives' (and thus of the War Office) about the 'radicals' was that they had a theory, and that was all. This theory was that the tank, and the tank alone, could strike hard and deep into the enemy's rear to decapitate an opponent's command functions. The theory had much to commend it, but it had never been wargamed with or by professional soldiers. In retrospect, the Second World War was to demonstrate that there was much good in these ideas, but that their extreme arguments were, in fact, false. The tank was indeed a critical part of a combined-arms force, along with mechanised infantry, self-propelled guns, ground attack aircraft, armoured reconnaissance and radio communication. Such combined-arms formations would be required to fight against main force enemy elements, not merely to threaten and disrupt their headquarters and command and control functions.

This was what the soldiers of the 1920s and 1930s were slowly groping for in the darkness. The War Office was right to take things cautiously, but the fact that the Experimental Mechanized Force was able to achieve only four days of formation exercises in two years (two days in each of 1927 and 1928) gave ammunition to those who criticised it for being half-hearted. The mistake was not to undertake these studies and trials rigorously enough, with a doctrinal template underpinning them, or to fund them adequately. At the end of 1928, for instance, Milne, an otherwise fair-minded CIGS, gave in to pressure from the General Officer Commanding Southern Command (Montgomery-Massingberd, another experienced and fair-minded man, whatever slander J.F.C. Fuller threw at him) to switch focus from a single 'tank' brigade trial in order to slowly mechanise the whole army – infantry divisions and cavalry brigades alike.* The Experimental Mechanized Force experiment was halted.

What Montgomery-Massingberd wanted was 'to use the newest weapons to improve the mobility and fire power of the old [infantry and cavalry] formations', rather than to create a new, specialised formation, even on a trial basis.[8] There was much sense in this. The problem was that stopping the Experimental Mechanized Force threw the baby out with the bathwater. The emphasis on developing the armoured brigades, in the process of which new doctrine might be presented and new opportunities grasped, halted completely. By stopping the Experimental Mechanized Force, the focus of the experiment shifted away from finding out how a divisional, corps and army commander might be able to *manoeuvre* on the battlefield, to one in which old-style troops had better *mobility*. Milne, who understood the difference between these two concepts (he had changed the name of the Experimental Mechanized Force in 1928 to Experimental Armoured Force to capture this distinction) nevertheless gave into Montgomery-Massingberd's pressure and ordered the end of the experiment. He did so, thinking he might have the opportunity to restart the Experimental Armoured Force after a year.

Unknown to Milne a financial whirlwind was about to hit the United Kingdom. In the financial depression which followed he found that he could not revive the Experimental Armoured

---

*The British Army at home was administered through six regional commands: Northern, Eastern, Western, Southern, Scottish and Northern Ireland.

Force. In 1929 and 1930 small-scale mechanisation trials in the infantry, cavalry and tank forces took place instead, examining a wide range of relatively low-level subjects such as the deployment of machine guns in the experimental infantry battalions, the use of radio communications between armoured vehicles, the employment of machine-gun carriers in the cavalry brigades and the role of light tanks within the infantry brigades. New and 'big' thinking about the creation of a modern British Army or of new tactical doctrine to prepare it for future war was limited once more to conversations between the small band of enthusiasts in the 'reformer' camp.

One product of the trials was the publication of what became known as the 'Purple Primer', named from the colour of its jacket. Lieutenant Colonel Charles Broad prepared his ideas for the organisation and deployment of 'Mechanized and Armoured Formations' which was published in March 1929. In 154 short pages (101 of which were given over to manning tables for various types of mechanised unit) it captured the practical lessons of both the Experimental Mechanized Force and the 1928 exercises. It is clear from these that a direct link was being made between mechanisation, mobility and the possibility for manoeuvre. For instance, Broad described the role of an 'Armoured Fighting Vehicle' to be to 'act by fire and movement with the immediate object of creating an opportunity for decisive action and the ultimate one of securing a concentration of superior force at the decisive point'. Broad asserted that by 'increased mobility the art of generalship is enhanced'. Likewise, the 'moral and material effect of A.F.V.s on other arms is great. They can, in fact, render immobile, by threat alone, such infantry formations as are unsuitably equipped.'[9]

There remained many gaps, not least in the concept of integrating infantry with armoured fighting vehicles, armoured supply, mobile armoured artillery and anti-tank weapons, or co-operation with air, but these concepts were a significant step forward in creating a combined-arms operational doctrine for the British Army. Broad recognised that he was drafting a document that would articulate the relationship between the fact of mechanisation and the possibilities for manoeuvre provided to the commander of the future:

> Victory ... has not always been obtained solely by direct attack. Superior manoeuvre may in the future, as in the past, cause the surrender of large and important forces, when the latter find themselves placed in impossible positions. A force may find itself in such a position if it is driven across neutral frontiers, cut off from the means of supply, or actually surrounded.[10]

It was, for its time, and following only a short period of limited experimentation, a remarkable volume, providing a foretaste of the development of battle during the next war. But pamphlets do not an army make. A concerted effort would be required to align the whole thinking of the army, together with its equipment procurement programme and its training effort to create the sort of force Broad envisaged. A revised pamphlet was issued in 1931, entitled 'Modern Formations'. It suggested the reorganisation of the army between mobile and non-mobile infantry divisions. The mobile divisions would be of three types. The first would comprise three horsed cavalry brigades. The second would have two tank brigades and a lorried infantry brigade, and the third would contain three tank brigades. The continued inclusion of the idea of horsed cavalry hinted at the slow progress being made in thinking of the future of warfighting in the army. The first chapter was given over to a description of the present stage of evolution of a new army organisation. It was unusual for such a document to comment so overtly on the reasons why the transformation of the army might not take place in time for the next war, but it reflected Milne's warning in 1927 about the absence of money:

> The decision to undertake the great expense involved in the rearmament of an army is ... likely to be delayed until the political situation indicates that an emergency involving serious

danger of war is within sight. It is more than possible ... that far-reaching changes in organization and equipment [and presumably, doctrine and tactics] may be delayed until the eve of mobilization or even until the outbreak of war itself ...

At the present time, when retrenchment and economy coincide with rapid development in all mechanical and electrical sciences, the gap between peace equipment and organization and our requirements for war is bound to be great.

These might have been an accurate reflection of the low status of questions in Britain of the army's preparedness for war – and it is clear from the context that Broad was referring to a new war in Europe – but it is nevertheless profoundly dispiriting to see the level of *official* acceptance (the Purple Primer was published under Milne's signature) to the *fact* of unpreparedness. By 1931 there was no money for the mechanised transformation of the army. Indeed, there was no money for anything. The value of exports (and thus the revenue they earned for British businesses and the tax take for the exchequer) had fallen by 30 per cent, unemployment rose rapidly from one to three million and a nadir was reached in terms of funding for the army, at £36 million. It was not politically possible for the army to receive money to develop tanks at a time when considerations were being given to reduce unemployment benefit.

The argument appears to be an acknowledgement that there was no point in even trying to wake Westminster up to the reality that the army was unprepared to go to war, because even with that knowledge there was no money to rectify the situation. Instead, the army was being told to live with unpreparedness and to do its best to undertake rapid transformation when and if it were forced to do so. In addition to being profoundly depressing, the argument failed to acknowledge just how important a role an effective and modern army could play in preventing war. An army is not merely a utilitarian tool, designed for a specific task. It can, if well conceived, play a role in political strategy as well, serving as a deterrent, for instance, the land equivalent of a naval 'fleet in being', designed to influence a potential adversary into a particular type of behaviour.

Broad's 'Purple Primers' were excellent demonstrations that the army, or elements of it at least, was beginning to understand what was required to transform it ready for a future, high-intensity war in Europe, or anywhere else for that matter, if it had the opportunity. They demonstrated that not all army officers were technical and tactical Neanderthals wholly opposed to any form of change to the way the army thought and behaved. Some were, of course, but most weren't. David Low's 'Colonel Blimp', when it emerged into Lord Beaverbrook's press in 1934, was an offensive caricature that created a public image of army officers as far from the truth as the 1980s *Blackadder Goes Forth* was about the Great War. Funny, but wrong. The documents were studied assiduously in Germany. But it was obvious, too, that much more needed to be done. Liddell Hart's quite reasonable though mild criticism of both documents was that they focused inordinately on structure and organisation, rather than on *purpose*. 'I would have liked to see the new manual deal more fully with the strategic potentiality of armoured forces for a long-range thrust into the enemy's rear, to cut his communications and arteries of supply' he wrote.[11]

One cannot help observing that the lack of money was an excuse for the lack of cerebral energy in both the army, which was paid to do such things, and in government, which was ultimately responsible for the issue of national defence, in at least thinking about how warfare would be fought in a future continental war. The issue was not so much the requirement to educate the army, but the need to educate the population as a whole, and its politicians. Major General Burnett-Stuart, one of the army's 'reformers', responsible for overseeing the Experimental Mechanized Force exercises on Salisbury Plain in 1927 and a man certain that war in Europe was coming, exchanged correspondence with Basil Liddell Hart when he was General Officer Commanding British Troops in Egypt in

1931, castigating Liddell Hart for constantly attacking the General Staff and thereby misdirecting his criticism:

> It is not the Higher Command that want educating – it is the public, the press and above all the Cabinet and Parliament. We [the army] know what we want quite well, but every financial and political obstacle is put in the way of getting it. Also, with a tiny army so dispersed as ours is, and always on duty, the practical difficulties of conversion are immense.[12]

The correspondence between the two men carried on into 1934. The nub was the army's ability and willingness to change itself to meet the requirements of a coming war. Was the army able to do this on its own without specific government direction? Liddell Hart suggested in reply that it was surprising to hear that if the army wanted to restructure itself, it couldn't do so because the government would object. Burnett-Stuart responded that this in fact was the case. The army could not simply decide to do what it believed to be right without specific direction from those who held the purse strings and determined the country's grand strategy. In practical terms the commitments the government placed on the army in respect of the defence of the empire, and the limited budgets by which it constrained what the army could or could not do (such as ordering the procurement of more tanks, for example) were very serious obstacles to freedom of movement.

The truth lay somewhere between Liddell Hart and Burnett-Stuart in this discussion. If it had believed in the necessity of reorganising itself for a coming European war (and the evidence is that it didn't), the army could have undertaken faster and more substantial change, setting out its design for a future, manoeuvre-based army in addition to the imperial police force it currently had. It could have undertaken some of this change of its own accord, such as, for example, designing a doctrine for this future army and taking radical steps to prepare this army to fight a future war in Europe, such as converting all of its cavalry divisions to armour. Then again, it would have required money to achieve this: the capital cost of a tank being far greater than a horse. Converting cavalry was easier said than done, because of the relationship the army had with India: India Command would need to agree any such changes because they impacted directly on Britain's agreement for the defence of India.

Indeed, the problem for the army was ensuring it could remain agile enough to manage the range of threats that the country faced, while encased in the Cardwell straitjacket. The hard link between the Home Army battalions and those in India meant that any attempt to reuse the Home Army battalions would upset the arrangement with India. A major war in Europe would mean the abandonment of Cardwell. The challenge for the War Office lay in determining the point at which the necessity for preparing for war in Europe meant the ditching of Cardwell. For instance, if an expeditionary force of five infantry divisions and a cavalry division of two brigades was created, it would require 60 infantry battalions from the 69 in the United Kingdom (59 line battalions and 10 Guards).* This would mean that 87 per cent of the Home Army would need to be deployed with the BEF, effectively ending the Cardwell arrangement.

Liddell Hart challenged the thinking about mechanisation, suggesting in a paper commissioned by the Cabinet in 1929 that the problem with it was that the army was yet to realise that modern warfare would be based on firepower rather than manpower. Scarce government money should be prioritised on forces that would give the greatest effect on the battlefield. On the basis of the last war this would be mobile, mechanised forces, not infantry and cavalry.[13] The War Office, however, had entered the decade fixed to the belief that it was numbers of fighting men that remained the essential demand of modern war, copying the huge numbers required between 1914 and 1918. In a public lecture in May 1933, the CIGS, Montgomery-Massingberd, argued this line, propounding that policing the empire did not require wholesale mechanisation,

---

*A further 69 battalions were based abroad.

as this was simply not required in the areas of the world where the army was expected to operate.[14] A report for the Committee of Imperial Defence in July 1932 reported, on the basis of the experience of the Great War, that in the event of another major war 750,000 recruits would be immediately required. Conscription would be a necessity, and that required money. The most important thing, therefore, was the rapid mobilisation of manpower, not mechanisation.

This thinking dominated conversations in the War Office about the transition to war through to 1939. The problem was again one of the baby and the bathwater. The issue wasn't of one thing or the other, but of both. The army needed to sort out its mechanisation policy in times of peace, so that when and if mobilisation happened, the newly recruited or conscripted soldiers could fit directly into the roles created for them. It was important for the army to prepare its plans, methods, structures, equipment and training for war during the times of peace. Waiting simply on a massive influx of manpower at the outbreak of war was not a plan at all. The evidence of these two decades, seen through both the Experimental Mechanized Force and the faltering steps towards mechanisation, was that the army paid lip service to the requirements for innovation and change.

Looking back, Liddell Hart said much during these decades that was helpful to the development of military theory. His concept of the indirect approach – the idea of sidestepping the enemy's strongest defences to attack its weaknesses, both geographically and psychologically – is as sound as it ever has been throughout human history. The two fundamental maxims of this idea were presented *inter alia* in his 1929 *Strategy*. The first is that:

> [in face of] the overwhelming evidence of history no general is justified in launching his troops to a direct attack upon an enemy firmly in position. The second, that instead of seeking to

upset the enemy's equilibrium by one's attack, it must be upset before a real attack is, or can be successfully launched.[15]

But, like Fuller, into whose shoes as chief military irritant he stepped in the 1930s, Liddell Hart's influence was also malign. Like Fuller's, his emphasis on the need to destroy the enemy's nerve-centre was, though theoretically sensible, sold as an alternative to the need for hard, confrontational battle. The emphasis on an all-tank army (and the relegation of infantry to a support arm) smacked of fantasy and unproven theory (and a desire to avoid the heavy infantry casualties of past wars), rather than on doctrine based on carefully observed experience and experiment. As the historian Corelli Barnett observed, 'When the British armoured divisions at last took the field in Africa in 1940–41 they suffered from this over-emphasis on the tank fostered by Fuller and his disciples.'[16]

Liddell Hart, on the other hand, saw operational manoeuvre as a combination of tanks, mechanised infantry and a tactical air force, very much as the Germans did when they created their first three panzer (tank) divisions in October 1935. Each division comprised two armoured brigades and an infantry brigade, all conforming to the same doctrine, training and command. The panzer brigades contained two panzer regiments. Each division, designed for deep penetration tasks, also contained a motorised artillery regiment and an anti-tank battalion, together with signal and engineer units. The panzer divisions were designed to be free-wheeling and self-contained, operating singly, together or in conjunction with less mobile formations, finding their way through the battlefield and beyond to create an operational (i.e. campaign-level) or even a strategic effect of the kind that had characterised the German March 1918 offensive and the subsequent, much more successful, Hundred Days battles.

For all the immensity of their contribution to the development of military science in the inter-war period the negative impact of both men was also profound. Both worked against the grain, rather than with it. Both saw enemies and opposition where there was none. Both fell into the trap of extremism, as well as on occasion fantasy. Both failed to take the room with them, alienating men who were happy to support and help develop rational ideas. Liddell Hart spent most of the 1930s attacking everything he didn't like from the side lines where he had the benefit of being able to criticise without the responsibility of delivering tasks made impossible by a lack of political will power, direction and funds.

The mistake was Liddell Hart's appointment as defence adviser to Leslie Hore-Belisha, Secretary of State for War following Duff Cooper's sideways move to the Admiralty in 1937. With the ear of Hore-Belisha, Liddell Hart set himself up in opposition to the Chiefs of Staff. In this role his undoubted brilliance in the arena of the creation of all-arms, armoured formations was undone by his determination to oppose a continental commitment and his involvement in the prosaic decisions of military life (such as promotions) which raised the ire of professional soldiers. Likewise, he took too long to realise that it was the structure of the Army around India (Cardwell) which would need to be replaced before the Home Army could be transformed. Perversely, his trenchant defence of limited liability for the Army in Europe prevented him from being able to use the prospect as the intellectual, doctrinal and practical basis for rebuilding the army along the lines he urged.

In the development of the Experimental Mechanized Force a handful of the 'radicals' – like Lieutenant Colonels Giffard Martel, Royal Engineers, and Percy Hobart, Royal Tank Regiment – held to Fuller and Liddell Hart's more extreme position with regard to building a tank-only force. Among the others, a variety of opinions were held, most of which supported the entire concept of a mixed-arms mechanised force. The creation of the Experimental Mechanized Force worked alongside experimentation in industry in respect of tank technology. Here again the absence of doctrine was to prove a handicap to the development of effective vehicles. Were tanks to exist to support the infantry in the attack or were they to take on the functions of the cavalry or, as Fuller and Liddell Hart advocated, independent shock action? The answer to these questions would determine the design of a suitable vehicle which was able to combine adequate protection, effective firepower and appropriate mobility.

The tank which they inherited was an excellent Vickers 12.5-ton Medium II Tank with a 3-pounder gun that could travel at 15 miles per hour. But it was weakly armoured and under-powered. Vickers proceeded to develop a replacement that weighed 16 tons. This, the A6, led to three Mark III prototypes being built but a new medium tank wasn't developed. The primary reason was a lack of money to develop and test effective alternatives. Prototypes are an expensive business and little spare cash was available. A compromise led to the development of a light tank (armed only with a Vickers machine gun) to fulfil most needs. Thus, a lack of cash, based on the Treasury accusation that the army didn't know what it wanted, meant that the British Army went through the entire decade without a medium tank at all.

Despite the lack of money, of an effective medium tank and of any combined-arms fighting doctrine, the army remained nevertheless committed to the principle of developing a concept of armoured warfare even though it was half-hearted in its execution. In 1933 the new CIGS – General Sir Archibald Montgomery-Massingberd – announced that the Royal Tank Corps would form a tank brigade. Its first commander was to be a 'radical' – Percy Hobart. Equipped only with light tanks they were, in absence of anything else, better than nothing. The five lost years of not having the Experimental Mechanized/Armoured Force could now be made up for by a fully operational brigade in the army's order of battle. The old problems remained – no medium tank in production and no combined-arms doctrine yet – but it was a start. Even better, in 1934 Montgomery-Massingberd authorised the creation of an armoured division in addition to the Tank Brigade, although again for reasons of economy this wasn't created until 1937 and was not ready until 1940.

The old problem of 'purpose' remained pertinent. Were the Tank Brigade and the new Armoured Division to be tank-only or combined-arms formations? The absence of clear doctrine upon which to build the new formations led to confusion and division, not least in respect of the appointment of its commander. A cavalryman might be expected for instance to develop the division along the lines of traditional cavalry doctrine (reconnaissance, flank protection and pursuit) whereas someone else might see a balanced mechanised force as a means to provide a corps commander with much-needed mobility in his conception of battle. The corps in which this division would operate would have only a single armoured division, the other two being infantry. These would be motorised (i.e. lorry-borne) though not mechanised, and therefore not able to fight alongside or integrated with the former, because they would lack its mobility and protection.

In addition to the issue of tanks and their deployment was that of mechanising the remainder of the army, especially that of the horsed cavalry. The solution arrived at was that the cavalry would gradually be re-equipped with light tanks or armoured cars and would serve either as light armoured brigades in the Armoured Division or as an armoured car regiment undertaking reconnaissance duties within an infantry division. Eight of the 55 horsed Yeomanry (i.e. part-time) regiments were likewise converted into armoured car companies and a further 20 into artillery regiments.

By 1939 there remained four horsed cavalry regiments in the regular British Army along with 15 Yeomanry regiments. Together these formed the 1st Cavalry Division, which was not much more than a strategic reserve of manpower, as it was deployed to Palestine in May 1940 to relieve infantry on internal security duties.* There was no doctrine for the deployment of a cavalry division in battle, which meant that the reality would be that, if required for battle, it would be broken into its various constituent parts and allocated across the deployed armoured and infantry divisions. There was also argument about whether or not the Tank Brigade should be included in the Mobile Division. It eventually was, in addition to two light armoured cavalry brigades. At the same time the Royal Tank Corps was being expanded in order to provide more battalions

*It was eventually mechanised in 1941 and formed the 10th Armoured Division.

for the direct support of infantry, several infantry battalions in the Territorial Army being converted also to battalions of the Royal Tank Corps in that role, but it is hard to escape the conclusion that the decade had been one of lost opportunities when the end of the 1920s had shown so much promise. Despite the Experimental Armoured Force, the Purple Primer and the decision to create an armoured division, and with the unequivocal evidence of a rapidly declining security situation in Europe, the British Army seemed less prepared than ever to fight a modern war on its doorstep. Was it just the British Army that struggled to divine the future or the country as a whole?

Chapter 11

Britain faces a rapidly changing world

The security of the world declined dramatically in the two decades following 1918. For the most part Britons were somnolently unaware of it. At home, there was next to no public opposition to the government policy of keeping the army starved of funds, and abroad the rapid development of a range of threats was dismissed by decision-makers as inconsequential either to Britain's security, or to that of its empire. At home the army was widely (and bizarrely) regarded by many as having been complicit in the bloodletting of the Great War. The argument, following in Lloyd George's slipstream, was that the generals were unnecessarily profligate with the lives of the nation's manhood and were not to be trusted with the management of war. In any case, so far as the readers of the newspapers were concerned, a new technological saviour had been divined – the bomber – that would keep any future European enemies at bay. Why, in these circumstances, should the country throw good money after bad by giving it to the army? It was the Royal Air Force that needed investment, as both the technology and the doctrine of air power were indicating.

If the followers of Liddell Hart had their way, Britain would never again have to accept what he considered to be an 'unlimited liability' for continental defence. In his determination to haul up the drawbridge around the English moat, Liddell Hart grievously ignored the single greatest lesson of the Great War: namely, that the physical security of the British Isles was intimately connected with that of Europe. One of the greatest deficiencies of British statesmanship during the inter-war period was the failure to grasp the truth that Britain was an essential component of the Western European security infrastructure. The doctrine of wishful thinking, assiduously advocated by Liddell Hart, was that the country could safely sit on the edge of Europe without committing itself to the possibility of war on the Continent. Instead it would continue to indulge itself in the fantasy that the greatest risk to Britannia was the myth of the Russian (or now, Soviet) hordes pouring into India, safely ignoring what was going on in Europe, and failing to prepare for the possibility of a re-run of 1914 given that the partition of Europe had sown old animosities anew. Britain had been grievously unprepared for the Great War. Would it be so unprepared for another? Sadly, yes.

The cause of this 'ostrichitis' was a failure of strategic vision, as much as it was of profound ignorance of the dynamics of power in Europe and its implications for Britain's security. This, of course, was a political failure, one of grand strategy. The 'British Way of Warfare', so the 'limited liability' theory postulated, was that the Royal Navy would continue to guard the sea lanes of communication across the empire, while the Royal Air Force, armed with war-deterring 'chemical bombs', would protect Britain's territorial integrity from 30,000 feet. The British Army, wearing solar topis, was consigned to deal with the empire.

This posed several problems for the army. At a political level it was hardly auspicious that the government considered, as did Neville Chamberlain when Chancellor of the Exchequer, that the Royal Navy and the Royal Air Force were the nation's strategic assets, because they were considered national deterrents, whereas the army was not. Nowhere in the annual surveys of imperial defence can be seen any argument made by the army's leaders of the concept of a modern, deployable army, operating on the precautionary principle (i.e. structured to counter a worst-case security threat) serving itself as a deterrent against bad behaviour in Europe. Likewise, it was hard to avoid inter-service rivalry with a new service loudly expounding on its strategic utility in respect of the defence of the country. This new theory of aerial warfare, based on fear, made the army redundant in considerations of national security, at least so far as home defence was concerned. It stood to reason that the limited pot of money available for defence had now to be shared three ways rather than two, with a technology (and research and development) hungry air force. The development of the bomber, the great white knight of Air Chief Marshal 'Boom' Trenchard's deterrent force, took first priority over the development of land technology, though it is fair to say that the Royal Air Force carefully and strongly defended its technological imperatives far better than did the army.* It was inevitable that some tensions were created, not least because of the zero-sum nature of the funding allocations. What was given to one could not be given to another. Hastings Ismay, when Assistant Secretary of the Committee of Imperial Defence after 1925, recorded that the sub-committee responsible for co-ordinating the defence efforts among the three services 'were signalised by the most blazing rows'.[1] Meetings were held as infrequently as possible, so as to limit the mutual hostility between the parties.

In Britain the decade and a half following the end of the Great War was one that was increasingly socially hostile to the concept of war. This sentiment was a natural follow-on from the trauma of war. Humankind was the stupidest of animals. It had lost so much in the war, and it might happen again. Faith in humanity's ability to order its affairs was low. Relief from the ordeal of war was found in the distractions of triviality. The mood was described by the Australian journalist Alan Moorehead:

The jazz era rose to its height, the world of bobbed hair and short skirts, of cocktails and cars, of cynicism and exhausted

---

*Trenchard was promoted to Marshal of the Royal Air Force in 1927 and retired as Chief of the Air Staff on 1 January 1930.

disillusionment, of the restless and hysterical clutching at little pleasures, and of dancing, dancing, dancing.[2]

The 1920s saw a general revulsion against the experiences of 1914–18 and the generals who, it was believed, had managed it badly. In many cases this took the form of anti-militarism and pacifism at one extreme and a belief at the other that war could be prevented by the creation of a Wilsonian 'rules-based international order' – to use that phrase before its time – made up of new global security organisations, the policies of which included such ideas as multilateral disarmament. The British Army undoubtedly felt the effect of this public disapproval. These antipathies built on society's long hostility to large standing armies. Armies were fine so long as they were garrisoned abroad and guarded the empire, not undertaking savagely expensive and inhuman endeavours of the kind that had been recently experienced in France. Accordingly, during this time the army focused on its core priorities and worried little about anything other than managing its diverse commitments. Its primary task was sustaining its garrisons abroad, not least in India. In 1928 the Chiefs of Staff third annual Review of Imperial Defence re-emphasised the imperial role of the army.[3] The primary commitments were:

1. Defence of ports, especially Singapore, Malta, Hong Kong and Gibraltar
2. India
3. Local security for Britain's direct interests abroad, e.g. Egypt, Sudan, China
4. The anti-aircraft defence of Britain
5. Garrisons for UK ports
6. Maintenance of United Kingdom's internal security.

The commitment to Palestine required an additional two battalions in 1929 following increased violence between Arabs and Jews, a responsibility which in the next decade seemed never-ending and unremunerative. Two battalions were still required in Sudan and in

China a military presence was required to protect British interests in Shanghai, Tientsin, Peking, Canton and Hankow, not to mention Singapore and Hong Kong. Driven in upon itself and not entirely unhappy in the cosy seclusion of its regimental life following its forced abandonment of the need to manage the complex mental and technical machinery of war after 1918, the army appeared, to those on the outside, as in many ways it was, antiquated and set in a mould that seemed more appropriate to the 18th than to the 20th century. The army had swiftly declined from national saviour to caricature.

All these influences meant that the armed services generally, and the army in particular, were the victims of a decline of public goodwill, which itself had an impact on public perceptions of the worth of military service, and of recruiting. A military career was regarded as one only for dullards. The Oxford Union famously voted in February 1933 by a margin of 275 votes to 153 'that this house will in no circumstances fight for its King and Country'.[4] The fact that most did so only a few years later didn't detract from the intensity of emotion against the Clausewitzian argument that military force was a legitimate tool of political action in a world where there was no intrinsic or systemic security. The same year the Fulham by-election resoundingly rejected a pro-defence Conservative candidate. The result was widely regarded to be a rejection of the argument for increased spending on defence.

The Great War had effected radical changes in society, from which the Royal Navy and the army seemed immune, bastions of a retrograde past. Only the Royal Air Force, and, in the army, the Royal Tank Corps and the Royal Army Service Corps as well as, in the navy, the Fleet Air Arm and the submarine service, were regarded in the popular imagination to have moved with the times. An army that in 1933 still had 136 infantry battalions and 17 horsed cavalry regiments, but only four tank battalions and two regiments of armoured cars, certainly gave that impression.

Pacifism played a very significant and growing role in the views of the British public during the 1920s. In 1923 the Labour member of parliament and future prime minister Major Clement Attlee reflected the views of many when he suggested to the House of Commons that 'the time has come when we ought to do away with all armies, and all wars'.[5] It was perhaps a piece of political theatre calling out to the very substantial pacifist lobby in the Labour Party. Attlee had been a member of the 'No More War' movement after 1918, proposing resolutions in the party to vote against the defence estimates. When in February 1924 he found himself Under-Secretary of State for War he immediately resigned his membership of Ramsay MacDonald's Union of Democratic Control – not overtly pacifistic but nevertheless advocating the reduction of military influence in parliament – and had by circumstance to change his tune. But his view reflected a very wide spectrum of opinion inside and outside of politics. In 1926 a majority at the Labour Party's annual conference carried a motion in favour of non-resistance. Pacificism became a very significant element within both the Church of England and the Labour Party, the two strands being reflected in the person and the 'Christian Pacifism' of Labour's George Lansbury when he took leadership of the Labour Party in 1931.

The exhaustion of war played on older strands of pacifism in the country, a movement that would grow steadily, reaching its apogee in the mid-1930s. Fear itself was a powerful ingredient in popular opinion. Since the early part of the previous decade the merchants of aerial death had been whipping up terror of the new type of air-delivered warfare in which 'the bomber would always get through', delivering death and destruction from on high, and bringing the battlefront to the home front.* Fear was fuelled by what turned out subsequently to be grossly inflated estimates suggesting that many hundreds of thousands of civilians would be

---

*The warning came from the prime minister, Stanley Baldwin, in the House of Commons on 10 November 1932.

killed by mass aerial attacks. This anxiety was reflected in popular literature, such as H.G. Wells' *The Shape of Things to Come*. In this new war it was probably safer to be on the front line than in defenceless British cities. The Trenchardian *idée fixe* was that a strong bomber force armed with chemical weapons (an idea strongly supported by Liddell Hart) could serve as a deterrent to war and therefore negate or reduce the need for other forces.

A year later the playwright, author and dedicated pacifist A.A. Milne, who had served in the British Army during the war (in fact, in the same regiment as Bill Slim and Bernard Montgomery, the Royal Warwickshire Regiment), wrote a book called *Peace With Honour* in which he suggested somewhat naively that war would simply disappear if humankind had the courage to renounce violence.[6] Many agreed with him. A raft of writers published novels in the late 1920s which derided war as primitive, barbarian and futile. It was an era of widespread public naivety about the reality of power in international relations. If enough people voted against war, so the belief supposed, war would itself become redundant. Unilateral disarmament (i.e. the giving up of one's weapons on one's own volition, regardless of the decisions of other parties, even of potential enemies) was a serious policy proposed ten times in parliament between 1924 and 1931. If one became less warlike, this argument suggested, potential threats would simply disappear into an ether of goodwill. This was a virus of wishful thinking of enormous potency and equally considerable naivety. The wider context of the social and familial trauma caused by the Great War needs to be taken into account when seeking to understand this folly. This was also a decade where thousands of the bereaved sought to reach out to their dead relatives in the spirit world by means of the occult.

But worldly-wise cynicism was fashionable too. With the world so seemingly out of control in 1925, T.E. Lawrence, no longer 'of Arabia' but serving as a private in the Royal Tank Corps under the pseudonym of 'Shaw', carved the Greek letters for 'Who Cares' ('*Ou phrontis*') on the lintel of his cottage. Lawrence understood the classical acceptance of power (and that of the irrational and chaotic intervention of the gods in the affairs of men) but he also knew of classicism's reverence for heroes. He, as 'of Arabia', had been one. But looking around him, now a humble private soldier, he recognised, as in Nebuchadnezzar's dream, that all such giants had feet of clay. World events could not be regulated outside of this chaos, so no matter how man planned, events would follow the course of those who had the power to direct them.

The power of pacificism became deeply ingrained in both society and politics. Ramsey MacDonald's Labour Party, returned to power in June 1929, campaigned that it would secure lasting peace on the basis of conciliation, negotiation and disarmament. The means of doing so was to be a renewed commitment to the League of Nations, an organisation which was expected by the democracies – not unreasonably – to be the forum where squabbling national antagonists could sort out their differences short of going to war. It did not represent the concept of collective and mutual defence of the post-war North Atlantic Treaty Organization (NATO) model. Clearly, the League of Nations would only be as effective as its members were serious about avoiding war. Collective defence requires nations to place their military forces at the disposal of a mutual agreement (again, like NATO after 1949) and it is here that the League failed. Its imperative was to secure peace through disarmament, whereas collective security requires the opposite.

Unfortunately, as it turned out the League became something of a mantra for people who were happy for their security to be paid for from someone else's bank account (i.e. by disarming) and with other people's blood. Indeed, the more talk there was of disarming (such as at the Disarmament Conference, which opened in February 1932), the less appetite there was in Britain to spend anything more than was absolutely necessary on rearmament. In any case, the Locarno Treaty in 1925 (committing Britain to come to the aid of Germany or France if either was the victim of aggression) and the Kellogg–Briand Pact in August 1928 (in which

65 leading countries declared breaches of national sovereignty to be illegal), appeared to create a sense in Britain that international agreements could prevent a future war. Locarno was one of those empty promises governments regularly make to their friends, fingers desperately crossed in the hope that the cheque would never be cashed. The idea of yet another commitment to the French was galling both to those who saw in Europe only the fruits of bloodshed and to those who saw Britain's defence to be focused on its empire.

The League of Nations was increasingly regarded as the guarantor of international peace, without the concomitant commitment by the democracies to remain armed against a threat from those countries which did not hold to such pacific ideals. Nation states were considered to be the originators of war; international law – despite the fact that it existed more in people's imaginations than in hard cold reality – was seen to be the solution. It was the Italian invasion of Abyssinia in 1935 that persuaded some of the Labour Party die-hards to transition from a 'no war at all costs' school to one that supported the collective security arrangements of the League of Nations (though still without Britain spending any more money on armaments). The National Government had a minister responsible for foreign affairs and a separate one in the Cabinet (Sir Anthony Eden) responsible for the League. No popular voice challenged the ten-year rule, first in 1919 and then again when it was reimposed in 1928, as its logic seemed unassailable. Until the mid-1930s there did not appear to be a threat in Europe, and Locarno, despite no British forces being allocated to it, appeared to wrap up any threats to France which might in turn demand a British commitment to the Continent, the traditional bugbear of the political and imperial warrior class.

The Locarno Treaty (the 'Pact of Mutual Guarantee') of 1925 formally committed the British Army to the promise of a continental commitment, though the means of delivering on this promise did not exist until the formation of the BEF in 1939. The problem was that Locarno reflected the reality of Britain's relationship with Europe, or at least, an attempt to preserve the *status quo ante*, as without it France might seek an accommodation with Germany, thereby nullifying everything that had been achieved in 1918. While Locarno guaranteed the Franco-German border it helpfully, for Britain, ruled out bilateral relationships between the signatories. It meant that Britain could 'commit' to French security without actually committing any troops to Europe. It was just the sort of deal that London liked: cheap.

However, as the 1930s beckoned the War Office grew increasingly worried about the continuing imposition of the ten-year rule and the financial stringency it imposed upon defence. In their Review of Imperial Defence in 1929 the Chiefs of Staff warned the Committee of Imperial Defence that at the first sign that the international situation was worsening, the financial strictures needed to be removed from defence, as 'the requisite preparations' for 'the necessary adjustments in our defensive arrangements ... would require a period of some years to bring into effective operation'.[7] The following year they again warned that 'This country is in a less favourable position to fulfil the Locarno guarantees than it was, without any written guarantee, to come to the assistance of France and Belgium in 1914.'[8]

The year 1929 saw the first cracks appear in the belief that Britain was protected from the prospect of future war by pacifistic wishful thinking at home and a naive belief in the power of collective security abroad. The crash of the American Stock Market at the end of the year dried up flows of capital investment into Europe, precipitating the Great Depression. In Germany in 1930 the National Socialist party increased the number of its seats in the Reichstag, making it the second largest party. The monetary crisis of 1931 dried up sources of finance for both private enterprise and government spending. In the elections of that year a divided Labour Party was removed from power, MacDonald returning to lead a 'national coalition' of his supporters together with the Conservative Party. The arrival of this coalition coincided with a dramatic change to the shape of the global security situation. These years were the historic fulcrum between the dramatic decline in military capability in the decade following the end of the Great

War, and a new era in which the realisation was slowly dawning that a new settlement was required.

The challenge for Britain was to understand what was going on in Europe and comprehend its implication for Britain's wider security, especially the challenges these changes made to Britain's accepted posture of imperial defence. For the most part these were years – until 1935 at least – in which British policy makers peered through the glass darkly, unwilling or reluctant to face up to the truth of what they could see. The principal roles were played by Stanley Baldwin, Lord President of the Council; Sir John Simon, Secretary of State for Foreign Affairs; and Neville Chamberlain, Chancellor of the Exchequer between 1931 and 1937. For his part Chamberlain was to become the whipping boy for the policies of appeasement that were to be a hallmark of his time later in the decade as prime minister.

The first shot across the bows of what passed for superficial global amity was in September 1931, when Japanese troops seized Mukden and then proceeded the following year to swallow the rest of Manchuria and create the puppet state of Manchukuo. The members of the League of Nations (which excluded the United States, which Congress had never endorsed) demanded that Japan withdraw but in the face of Japanese recalcitrance were unable to decide on a response, with the result that China's appeal to the world for action to protect it from aggression fell on deaf ears. Hand-wringing by members of the League who, when faced with the brutality and finality of actual power found themselves unable to enforce compliance of a member with rules made in an ivory tower, led to their having to back down and back away. Japan withdrew from the League in 1933.

The affair exposed the entire naivety of a global project with only 31 members and no practical collective unity. On the one hand Japan had violated China's sovereignty and had opened itself up to recourse to the economic sanctions provided by the Covenant of the League of Nations. On the other, Britain was unwilling to provoke Japan's ire, even if it could secure the support of Australia and New Zealand for sanctions. Britain then proceeded to anger the United Sates by refusing to support Secretary of State Stimson's pronouncement on 7 January 1932 of 'non-recognition' of Japan's illegal acquisition of Manchuria. In any case, appeasing Japan was considered by many in London – including Neville Chamberlain – to be a sensible way of containing a potential problem in a country a long way away and between people of whom we knew nothing, to coin a phrase Chamberlain was to use in 1939 in relation to Czechoslovakia.[9] The long-term impact of such a policy in respect of emboldening Japan was not considered. For Britain, Japan was far away, and Manchuria was meaningless to British interests. Or so it thought. Few people in Britain were honest enough to accept that the Manchurian affair demonstrated that the League of Nations, long regarded by the naive as the bulwark against nationalist aggrandisement, was but a paper tiger. It would take several more years for the penny to drop.

The period between the crash of the financial markets in 1931 and the rise of Nazism in Germany in 1933 was one in which governments in London were increasingly concerned about the need to increase defence expenditure, while being politically reluctant to do so for two reasons. The first was the lack of political appetite for defence spending, reflecting the very significant degree of pacifism across the electorate. If politicians voted for defence spending, they knew they were likely to be voted out of office as the Fulham by-election had seemed to demonstrate. Secondly, the financial crisis made spending extremely difficult. Indeed, the Treasury believed that over-spending was a greater risk than war, opposing the removal of the ten-year rule and the opportunity this would provide to rearm. In the debate to end the ten-year rule the Treasury refused to accept the argument for rearmament, agreeing only that the case for additional spending needed to be made on an exceptional basis, with due regard to the dire financial situation the country faced.

In January 1933 events closer to home were to change the security landscape dramatically, with the seizure of power by the Nazis in Germany. Before long Germany's new leader, Adolf Hitler, announced that Germany would exit both the Disarmament

Conference and the League of Nations. The writing, as the Committee of Imperial Defence stated, 'was on the wall'. Like Belshazzar, however, the language was one that Britain did not either fully comprehend or take seriously. Most people did not consider that the rise of a new dictatorship in Germany meant that war was inevitable. In any case, Britain's economic woes took priority over any rethinking of the needs of defence. Nevertheless, the Cabinet in 1932 reluctantly accepted the representations of the Chiefs of Staff to cancel the ten-year rule.[10]

In the years following the Chiefs of Staff February 1932 Review of Imperial Defence a number of Cassandras began to worry, in public and in private, that the war clouds were looming. The CIGS, George Milne, in a remarkably prescient memorandum in October 1932, warned the Cabinet that if German foreign policy continued its current trend the inevitable result would be that when it was ready it would attack Poland, and then turn on France.[11] He wasn't advocating getting involved in these squabbles, however, merely warning of them. Likewise the Review warned the Cabinet of the weaknesses of Britain's defences, especially in the Far East, suggesting that the ten-year rule was wanton foolishness when 'war might actually begin tomorrow'. His successor, however, Field Marshal Sir Archibald Montgomery-Massingberd, only a few months later reinforced in a public lecture the primary role of the army to defend the empire, rejecting the idea that it was time to reorganise the army for a major war in Europe.[12]

The following year, the Chiefs of Staff Review of Imperial Defence for 1933,[13] submitted to the Cabinet that October, made it clear that the commitment to imperial defence had made the army utterly unable to respond to a crisis on the Continent. At the most, two infantry divisions could be provided as an expeditionary force, half what was initially deployed in 1914. It was the opening salvo in a fight in Cabinet between the Chiefs of Staff and the Treasury to create an effective expeditionary force available for deployment to the Continent in the event of war. A month later the Cabinet confirmed that defence expenditure should be prioritised on the three-fold basis of the defence of British interests in East Asia, commitments to the Continent and the defence of India. A joint committee of the Foreign Office and the Defence Requirements Committee was established to determine where the greatest deficiencies lay across all three services. It wasn't to build new capabilities, but to fill empty ones.

The report, presented in February 1934, painted a horrifying picture of deficiencies across the entire defence environment, and recommended a radical programme of financial and material rectification for gaps in capabilities that had been allowed to grow over the previous decade. At the same time the new regime in Germany formally placed the country in the War Office's sights for the first time as a threat to Britain's security in Europe.[14] The core assumptions were three-fold. The first was that 'we take Germany as the ultimate potential enemy against whom our "long-range" defence policy must be directed' (described as 'The German Menace'). The second was that Japan was a threat that needed to be ameliorated by a 'policy of accommodation and friendship' and the third was that India remained the largest of Britain's imperial commitments.

The greatest military deficiency the committee identified in the army lay in the weakness of the forces allocated to a future expeditionary force able to be deployed abroad. A second BEF (i.e. following the first one, in 1914) would be essential to protect the independence of the Low Countries which, if in the hands of a hostile power, could bring the whole 'of the Midlands and North of England ... within the area of penetration of hostile air attacks'. To be effective, once the first division had been deployed, it would need to be increased from two to six divisions over a period of six months. Even a six-division commitment would be completely inadequate, the report argued, as bitter experience in 1914 had demonstrated.

The report fired the opening shots in a campaign to ensure that Britain fund a substantially enlarged expeditionary force far beyond the two infantry divisions that were currently allocated for this task. It argued that Britain needed to prepare an expeditionary force within a month of the outset of war that totalled four regular army infantry divisions, a cavalry division, two air-defence brigades, a tank brigade and ancillary services together with three months' worth of ammunition for combat operations and sufficient reserves. It would need to be supported by 19 squadrons of aircraft. This was the 'absolute minimum' of what would be required if it were to be effective. Subsequent contingents would have to be drawn from the Territorial Army, which would need to be specially designated as available not merely for home defence purposes. The cost of building the expeditionary force to this level (i.e. four infantry divisions and one cavalry division) was estimated to be £25,680,000, a drop in the ocean relative to the entire defence-wide shortfall estimated to be £132,498,180 over five years.

In making the case for an enlarged expeditionary force the committee acknowledged that any commitment to deploy the Territorial Army abroad would require that it be appropriately trained and equipped, to a level equivalent of the regular army. The cost implications of such a course of action worried the War Office and Treasury equally and were part of the reason for their tardiness over the years which followed to agree to the committee's recommendations. Additionally, the logic of agreeing to the deployment of the Territorial Army abroad was that, to replace these troops in due course of time, conscription would be required. This was a slippery slope in terms of a commitment to war that simply wasn't acceptable until events in 1939 made it unavoidable.

The report was exposed to the full furnace of Cabinet consideration in March 1934. All involved recognised that the acceptance of even a small part of it entailed a significant change in government policy. Despite considerable reluctance being voiced by the Chancellor of the Exchequer, Neville Chamberlain, the Cabinet agreed in principle to the recommendations of the Chiefs of Staff in respect of the provisioning for an expeditionary force.

Chamberlain detested with every fibre of his being the idea of another war. He wasn't a pacifist, but – despite not having served in France – he hated the idea of another as destructive as the last.[15] He had a visceral detestation of those in the other camp – which he unfairly branded the 'arms lobby' – who argued the opposite case. But he also recognised, in 1934 and 1935, the reality of the rapidly declining international situation and the threat this posed to Britain. He also understood the need for strong armed forces in respect of negotiating with a newly emboldened Germany. Accordingly, as Chancellor of the Exchequer, he pressed for the costly rearmament of parts of Britain's defence, primarily those associated with air power. He strongly opposed any argument to increase substantially money for defence.

Chamberlain strongly opposed the argument for an expeditionary force and the money needed to create the force, in the midst of the current severe economic depression. His commitment to balancing the country's books meant that he could not conceive of the possibility that Britain might be forced into a war without the (financial) means to fight it. At the same time the Treasury was firmly of the view that increasing expenditure on the army would worsen the nation's finances and that the best way to defend the country was to get its economy on a sound footing before indulging in any more expenditure. In any case, if money were spent on the Royal Air Force or Royal Navy it would at least help British manufacturing, but expenditure on the army was seen as a poor investment. Cinderella was most certainly not invited to the Treasury's ball.

When the Defence Requirements Committee's proposals were presented to Parliament by Duff Cooper (by now Financial Secretary to the War Office), on 17 March 1934, the enhancement to the expeditionary force was justified on the basis that it was a development of the existing model and could be deployed anywhere in the world where it was required. An expeditionary force was not purposed solely for another European war. Duff Cooper's reluctance to make express connection with a commitment to Europe, and that of the War Office, was based on their hard-headed acknowledgement that both public and parliamentary opinion was still firmly set against the idea of another European war. Even the title – expeditionary force – was a red rag to some. But the government pressed ahead with a White Paper on rearmament a year later, using as the rationale Britain's commitment to meet its treaty obligations, the principles of collective security and the League of Nations. To the demand to 'stand up to the dictators' came the reasonable riposte 'with what?'

For a very significant proportion of the British electorate this dichotomy was never resolved: they wanted their collective security cake, with 'The League of Nations' printed on the icing but were reluctant to contemplate building up the armed forces that would contribute to that security. Many fervently disbelieved the proposition *si vis pacem, para bellum,* considering this to be a militarist's charter and the whole idea of using force to protect democracy to be a contradiction in terms. The arguments presented by Lieutenant Colonel Hermann de Watteville in 1919 were rejected outright. But there was no avoiding the reality that the desired collective security – through the League of Nations or elsewhere – was meaningless if each of the parties to the agreement were unprepared to underwrite their commitments with adequate military force.

The Defence Requirements Committee programme estimated the cost of its first assessment of military deficiencies to be £75 million over five years. Chamberlain, responding for the Treasury, argued that this was too great a sum for the country's finances to bear. Priorities would have to be made. A long campaign in the press, promoted assiduously by the now retired Trenchard, suggested that the most economical way of defending the British Isles was by using air power rather than by creating a larger or more deployable army. Chamberlain, in any case horrified at the thought of another war that mirrored the last, was an easy convert to this beguiling but ultimately foolish idea. A fascinating aspect of Chamberlain's response to Duff Cooper's presentation of the Defence Requirements Committee's proposal was the Treasury's usurpation of the role of strategic adviser to the Cabinet.

Chamberlain's response laid out a series of arguments as to how the next war would be fought, as justification for his recommendation that the Royal Air Force, rather than the army, should receive the bulk of any new investment. First, the Japanese would need to be allocated the lowest priority, as it was German rearmament that the public were afraid of. Secondly, threats of a German attack against France were overblown given that France now had the Maginot Line. The Germans would therefore attack Belgium instead, which would be overcome by massive Luftwaffe superiority, making any British force in the Low Countries a waste of effort. Instead, the threat of German air attack would require a strong air force based in Britain.

Chamberlain's rearmament proposals were, instead of supporting the proposal for an expeditionary force, to increase the Royal Air Force from the 1923 programme of 52 squadrons (which was still ten squadrons short in 1934) by a further 38. If attacked, the Royal Air Force would engage and defeat the intruders while the army used this time to prepare to cross the Channel to protect the Low Countries. The subtext was a reluctance to countenance creating another BEF, a consequence of which would inevitably be a call for conscription. If there was anything so guaranteed to put a wrecking ball through the economy, it was compulsory military service.

Chamberlain's response, demonstrating his political ascendancy over the Chiefs of Staff, challenged their prerogative to determine how an enemy attack on Britain would be conducted. The response by the Chiefs of Staff to this criticism was surprisingly muted, although privately they fumed. Colonel Henry Pownall, then on the secretariat of the Committee of Imperial Defence, scribbled in his diary that Chamberlain's ideas on strategy 'would disgrace a board school'.[16] In any case the usurpation of the advisory role of the Chiefs was remarkable, and calamitous to their rational exposition of the country's security deficiencies. Politically, Chamberlain's approach was expedient, though future events were to demonstrate the foolhardiness of his railroading of his Chiefs' advice. The idea of Trenchardian air offensive against Britain as the sum of an enemy attack on Britain, on which Chamberlain had hung his hat, was merely a theory: a strategic attack on an undefended British civilian population outside of general war and as a principal war aim would cross ethical boundaries that even Germany at the time would not countenance. In any case, a German attack on Britain would never constitute an air campaign alone, but would involve combined land and air attacks, the first of which would be a land offensive to seize the Low Countries (and its airfields) from which air attacks on Britain would then be carried out. The CIGS, Montgomery-Massingberd, suggested – correctly as it transpired – that the Luftwaffe's priority would be the tactical support of the ground armies and not the much-worried-about aerial bombardments presented in lurid details to their readers by the popular press.

In the end Chamberlain's political ascendancy meant that he won this battle. The Defence Requirements Committee plan was reduced, and the army allocated a mere £20 million over the five-year period. The bulk of monies went to the Royal Air Force for the air defence of the United Kingdom. This wasn't necessarily a bad thing if the right aircraft were purchased and if the army was also adequately funded. Neither proved to be the case.[17] So far as the Treasury was concerned the argument for an expeditionary force remained unmade, although the Chiefs continued to stand by their principles in the years ahead, arguing consistently that Britain's defence posture required a BEF, despite its cost or political unpalatability.

During the entire decade which led up to war in 1939, and despite Chamberlain's rearmament programme and the fear of aerial bombing exacerbated by Guernica in 1937, there was no public drive to rearm. No public bodies, organisations, lobby groups or newspapers outside of military circles publicly advocated the *si vis pacem, para bellum* principle. It was left to the Chiefs of Staff to make all the running – intellectual and practical – to ensure that Britain and its national interest remained defended. In the detail of the White Paper in March 1935 were express references to the withdrawal by Germany and Japan from the Disarmament Conference and the League of Nations, and Germany's decision that year to rearm. In such circumstances Britain's commitment to the Locarno Treaty would be meaningless without the means to enforce its treaty obligations.

The debate in Parliament was not about initiating an arms race, but merely making up for some longstanding deficiencies, especially in building up the Royal Air Force. This began with announcements to increase the number of aircraft in the Royal Air Force, from 850 to 890 announced in March 1934, to an announcement in July to go further, taking the total at home and abroad to 1,304. Political opposition was largely driven down party political lines. The Liberals worried that it represented a new arms race and Labour remained opposed to anything that wasn't disarmament. It was the short, ten-page White Paper on 1 March 1935 that finally reversed the policy of the previous 16 years. Britain must put its 'defences in order', it concluded, stating that additional 'expenditure on the armaments of the three Defence Services can, therefore, no longer be safely postponed'.[18] Despite this, when Ramsay MacDonald, the Labour leader of the National Government resigned in June 1935 and was replaced by Stanley Baldwin, the new government was elected in part on Baldwin's promise of 'no great armaments'.

The realities of the rapidly changing world did not persuade the public to change its mind about the need to prepare for war. The Great War was still too close in the national consciousness for efforts to prepare for another war to be taken lightly. The opposite in fact happened. The pacifist campaign was strengthened by the creation of the Peace Pledge Union by Canon Dick Shephard in 1934, a movement that folded within its arms a number of earlier groups. Separately in 1934 and 1935 the League of Nations Union, a large and influential part of the British peace movement, organised a house-to-house canvass across the country in which over 11 million people took part. This was over half of the electorate. Over 10 million people voted that Britain should remain a member of the League, for negotiated all-round disarmament and for the abolition of the private manufacture of arms. Over 80 per cent voted for the abolition of all national air forces. On the crucial question of sanctions, 85 per cent voted in favour of economic sanctions and 74 per cent in favour of military measures if necessary.[19] The challenge for the government was to ensure the security of Britain's interests while at the same time keeping an eye on the very significant views of the electorate with respect to defence.

The first signs of the unravelling of the post-Locarno and League of Nations nirvana came with the demand of Germany at the League of Nation's World Disarmament Conference in Geneva in 1932 for what might be called 'parity of esteem'.* It hardly seemed fair, it argued, that Germany should be held to the provisions of Versailles when making disarmament decisions. Britain had long been irritated at French insistence of its security above all else, some in Westminster arguing that French security guarantees were in fact the best way of ensuring the onset of another war. Indeed, the belief was strong among many, in government and without, that insistence on unilateral security was an intrinsic threat to the security of the group.

The truth was, however, that in Europe in the early 1930s old and new animosities were slowly being woven into a new family of threats that posed a dramatic challenge to British security. In Italy, Mussolini started his erratic course to rebuild a Roman empire. In Germany, from the start of the Third Reich in 1933, the Nazis began secretly to work against the obligations of Versailles. And Russia remained a problem, not just in terms of the Great Game in Asia, a worry about which the British seemed reluctant to relinquish. For the years which followed the range of potential threats, always presciently articulated in the annual Chiefs of Staff Reviews of Imperial Defence, served to create a nightmare of conflicting challenges for a cash-strapped Cabinet. Which way to turn? With few other practical options, it was determined to kill

---

*Hitler removed Germany from the Conference after he came to power in 1933.

Japan with diplomatic kindness, and to deter Germany by a rapid build-up of bombers for the Royal Air Force.

In the years after 1935, while it could be said unequivocally that Germany was increasingly powerful militarily, few believed that it posed a direct threat to Britain. There was much pro-German sympathy in Britain (though much less for Italy). The left blamed Versailles for Germany's political and economic woes while the right not so quietly hankered after some of its socially disciplinarian tendencies. Most politicians and administrators in Britain were prepared to give the new regime in Germany the benefit of the doubt. Few had the fierce Germanophobia of Sir Robert Vansittart, permanent Under-Secretary of State for Foreign Affairs at the Foreign Office between 1930 and 1937, who saw German-ness as the root of all evil.

PART THREE

New threats and new realities

Chapter 12

Boiling the frog: the rise of the Nazi threat

From the time of his seizure of power in Germany on 30 January 1933, the vast majority of Britain's politicians and pundits got Hitler wrong. Most admired him for making the trains run on time, and were impressed by his ability to re-energise Germany from the dreary days of Weimar depression, but simultaneously grievously underestimated the threat Nazism posed to Europe. During the nearly seven years before war broke out again in Europe on 3 September 1939 (noting that the Second World War began in China in July 1937), Britain's delusions with respect to its international security boiled – like the proverbial, unsuspecting frog – merrily away. Only a tiny few – men like Winston Churchill and Sir Robert Vansittart – heeded British and American reporters in Germany (few British diplomats actively rang the alarm bell) who maintained a steady stream of warnings from the very start of the Nazi regime.

In *The Times* on 31 January 1933 the journalist Matthew Halton, reporting from Berlin, described the previous night's events in the nation's capital, when the Nazi Party under Adolf Hitler had taken control of the Reichstag. A man now on a crusade against the new totalitarianism taking root in Germany, Halton pulled no punches, using language that made many in the West consider him one of those alarmists who naively saw the world through a simplistic, moralistic and sensationalist lens:

> Strutting up the Chancellery steps is Hitler, the cruel and cunning megalomaniac who owes his triumph to his dynamic diabolism and his knowledge of the brutal corners of the human soul. Surrounding him is his camarilla of braves: the murderous fat Goering, a vain but able man; the satanic devil's advocate, Goebbels; the cold and inhuman executioner, Himmler; the robustious radical, Roehm, organizer of the Brown Shirts ... Their supporters were decreasing in numbers; but by intrigue and a trick, Germany is theirs. Supporting this terrible elite are the brown-shirted malcontents of the S.A., nearly a million of them, and the black-uniformed bullies of the S.S., the praetorian guard. And the whole structure is built on the base of a nation whose people are easily moved by romantic imperialism, by the old pan-Germanism, in new and more dynamic form, by dreams of Weltmacht [World Power] and desire for revenge, and even by nostalgia for the jungle and for the tom-toms of the tribe.[1]

The language Halton had begun to employ when describing what he was seeing around him during his visits was urgent, full of warning for what would happen to Europe, and the world, if it ignored what was happening at great speed to Germany. Laying aside journalistic convention, he became one of a tiny group of prophets who began insistently to warn the world of impending doom. 'The things I saw being taught and believed everywhere in that nation,' he wrote in one of his reports from later in 1933, 'the superiority of one race and its destiny to rule – will one day become the intimate concern of all of us.'

Halton was in no doubt about what Hitler's ascent to power meant for Germany. He visited twice that year, in January and then, more extensively, in the spring. From the very first instant he suspected that the name Hitler had given to his political party – the National Socialist German Workers' Party – was an enormous, brazen hoax, designed for one end, and one end alone: an aggrandising, racist agenda that would attempt to place Germany ahead of every other country of Europe, with force if necessary. From what he could see, Hitlerism was neither national nor socialist, nor, for that matter, about the workers. It seemed to him to be about the creation of a racially pure Germany (in which Jews, Gypsies, and Slavs were obvious disposable imperfections) that was disciplined, obedient, militaristic and imperial.

Most worrying of all was that the fabulous lie inherent in the banality of the name of this political party was widely believed, and constituted the basis of Hitler's ascent to tyranny. To the fanatical few, the believers, the National Socialist German Workers' Party was a means to an end, the first step of which was the assumption of total, dictatorial power. The principal themes of the 24 February 1920 Declaration – racial purification from within and territorial aggrandisement outside to form a *Grossdeutschland* ('Greater Germany') of 80 million Germans, in which *Lebensraum* ('living space') – would then be achieved, with legal violence if necessary. To those who were prepared to understand the Nazi programme, the union of all Germans was an article of faith and had to be achieved, even if it entailed the spilling of blood. This was the Nazis' destiny. It was, too, Hitler's personal mission. This is why fate had called him to the highest office in the German state.

The name therefore fooled lots of Germans, as well as much of the watching world, few of whom ever bothered to read (or knew where to look for) the defining articles of Nazi faith. The lie was carefully concealed within other myths of Germanhood, ones that enjoyed a widespread appeal, particularly among a people humiliated by defeat in 1918, confused by the political uncertainties of the Weimar years, and pauperised by the collapse of Wall Street in 1929. It was the latter, much more than the impact of 1918, that lit the path for Germans, and Germany, down the road to the gates of a Nazi Valhalla. With a shock Halton understood that the Nazi creed was not being foisted unwillingly upon a reluctant population. Many in Germany welcomed the Nazis with open arms. 'Germany enters a nightmare,' Halton wrote. 'I feel it in my bones. She has heard the call of the wild.'

In Britain, however, most opinion – public and political – dismissed Hitler and his goose-stepping acolytes as goons and Halton's journalism as embarrassing hyperbole. In the comfort of his English home, the popular writer H.G. Wells dismissed the Nazi *coup d'état* that placed Hitler on the chancellor's chair as a 'revolt of the clumsy lout', a charge that led to the Nazi burning of his books – among thousands of others – in Berlin's Opernplatz on 10 May 1933. Wells underestimated the simplicity of the Nazi programme, the ferocious tenacity of its adherents, and the unchallenging acquiescence of most of the population. His accusation proved to be naively offhand, akin to the view among the aristocracy in the Heer (i.e. the German Army: together, the German armed forces were known as the Wehrmacht) that they could control the 'little corporal'. Wells gave no thought to the consequences that the rise to power of the lout he described would have for Germany, or the world for that matter. This was no schoolyard bullying that would disappear with the maturity of age.

That Hitlerism was ever able to so dominate European and global politics to the extent that it did for 12 hellish years, sending the world screeching into a cataclysmic war, is no mystery now. In the early 1930s, however, the fast-approaching catastrophe just wasn't obvious to everyone. Even though Europe was awash with journalists reporting back fearfully about the developments in the heart of Europe, where the Enlightenment legacy of centuries – before which lay the Reformation and before that the Renaissance – was being buried by a darkness so unfathomable as to be unimaginable, few read the warning signs correctly. That the warning bells were ringing, loudly, in every country of Europe and indeed across the world is indisputable; what seems so shocking is that few bothered to lift their heads to listen.

Another of those who, like Michael Halton, noisily rang the alarm bells, and another who was roundly ignored, was the journalist Leland Stowe, whose fears were succinctly articulated in a small book published in London in December 1933 with the uncompromising title *Nazi Germany Means War*.[2] Stowe, who had received a Pulitzer Prize in 1930 for his reporting of Great

War reparations, was as shocked as Halton after spending two months in Germany that year. He concluded that Germany had two voices. One was a public voice meant to pacify the fears of outsiders and spoke repeatedly of peace. The other was for internal, German consumption, and spoke relentlessly of martial values, social discipline, the importance of the needs of the state over those of the individual, Germany's requirement for living space, self-evident German racial superiority among the nations of Europe, and the imperative to achieve a homeland for all Germans, not just those who currently had the good fortune to live within the present boundaries of the Reich. What this meant for Austria, and for large slices of Poland and Czechoslovakia, was clear to the Nazi propagandists, and a message preached diligently every day in Germany. These ends would be achieved, Hitler asserted in *Mein Kampf*, 'by a strong and smiting sword'. 'Hitler declares that Nazi Germany wants peace at the very moment when Nazi Germany is busy, with an appalling systemization and efficiency, preparing its 65 million people for perfected martial co-ordination such as has never existed before', Stowe warned. 'This, in its logical sequence, can finally lead only to war.'

Stowe's book flopped. Neither buying public nor politicians wanted to spend money on a tome that suggested that on its current course war was inevitable and confirmed that their persistent 'ostrichitis' was a terminal illness. Equally, Halton's reporting of his two-month tour of Germany in the autumn of 1933 was regarded to be so alarmist that he was dismissed by many as a warmonger. He was criticised for reading into situations and events a meaning far beyond their reality. For suggesting, for example, that the Nazis and their followers were not Christians, truly following the sayings and precepts of Christ, he incurred the wrath of Catholics and Protestants alike, who accused him of ignorance. Germany was, after all, in terms of church attendance, the most observant country in Europe. It was factually erroneous, he was told, to suggest otherwise. As it was in Britain, pacifism was the dominant theological worldview across western Christendom in the 1930s. Although it attempted to express in political terms Christ's blessings on the peacemakers and the instruction to 'turn the other cheek', it seemed to assume that simply believing in non-violence would somehow persuade an enemy to think twice about using violence at all. Pacifism was the spiritual dimension of the political delusion of wishful thinking. As one critic observed in 1935:

> It may be comforting to consider that there will not be another war, but it is very foolish to do so, if only because such an attitude justifies that lack of preparation which serves as an encouragement to an aggressor ... The abolition of fire brigades would not mean the end of fires, nor that of knives, forks and spoons that appetite would disappear, nor that of hospitals or doctors that disease would in future leave us free. Neither by treaty nor by law could we secure ourselves entirely against any one of these.'[3]

Within Germany Halton, and others like him, were accused of not understanding the country or of respecting the legitimate German hunger for a new political settlement and an honourable place in the world. It was a common enough accusation. Although Halton robustly dismissed the Western reactions to his warnings, he nevertheless worried at the time that well-meaning ignorance of Nazism abroad was almost as bad as Nazism itself. If people only knew what hatreds lay at the heart of the Nazi creed, he thought, they would oppose it just as strongly. But because Nazism was cloaked in a cunning disguise, few people could see it for what it was.

Indeed, he believed that the fox was already inside the chicken coop. The German people, or at least large numbers of them, had embraced ideas they would have regarded, in another political or cultural milieu, as desperately irrational. Vast swathes of Germany now espoused nonsensical racial views about their own superiority over other varieties of humankind that had no place in rational or scientific thought and which, as Halton observed, 'one would have expected children to laugh at'. What had happened to the most intellectual country in Europe? He concluded despondently that it seemed apparent 'that the Germans were the least intelligent, if the most intellectual, of Western peoples'. Using the same analogy, they were also the most religious, if unchristian. Shockingly, the crazy notions of Aryan supremacy propagated so assiduously by the Nazis had already received academic and intellectual legitimacy and had been translated into notions that had quickly become widely accepted by otherwise thinking people, taught in schools and subsumed into common, everyday thought. He noted the prolifically published arguments of men such as geographer Professor Ewald Banse, whose *Military Science: An Introduction to a New National Science* published in 1933 argued among other things that:

War is both inevitable and necessary, and therefore it is imperative, and the nation's mind must be directed towards it from childhood. Children must learn to infect the enemies' drinking water with typhoid bacilli and to spread plague with infected rats. They must learn military tactics from the birds, hills and streams.

How was it possible for the most cultured society in the world to embrace such extremism? Why – and how – could otherwise deeply intelligent, well-educated, rational men and women embrace such nonsense? Halton observed the transformation of the German mind at first hand. The recipe was simple. If one lived within a lie for long enough, it didn't take long to fail to distinguish the lie from the truth. Ultimately, the power of the lie would trump reason and the exercise of rational thought. Indeed, one began to believe the lie. In 1930 he had made several friends in Germany during visits when a student at the University of London. His new acquaintances were all socialists and internationalists and laughed at the Nazi buffoons clowning around on the outskirts of politics. Three years later, after Hitler's rise to power, he went to meet one of these men, living with his parents in the Rhineland town of Bonn. Halton was relieved to learn that this man had not supped from the Nazi cup but was disquieted to hear that two of the others had become Nazis. Five years later, he returned to the family home during the Bad Godesberg conference in September 1938. 'What a pity you should have come up from Godesberg,' the man's mother exclaimed when she saw Halton at the door, 'because Friedrich is there! He commands a detachment of the S.A. which was sent from here to the conference.' All that was required was time, and the repeated articulation of the lie.

In 1933 Stowe commented on the catchy little musical ditty played during intermissions in radio programmes, which proved to be the popular song '*Volk an's Gewehr*' (People, to Arms!). On the streets, young boys dressed in Hitler Youth uniforms played with wooden cannon in the parks and practised throwing hand grenades as part of their school curriculum. At the time Halton, among many others, was concerned about the vast gulf that existed between what the Nazis said they were doing and the interpretation of these things by newspapers, politicians, and observers in other, far distant places, especially those safely cushioned in the protective cocoon of Western democracy, who thought that all systems of government operated similarly, and where bad people were constrained by the law and the structures and systems of civilisation.

But it wasn't only wishful-thinking Western intellectuals and politicians who harboured delusions about the Nazi programme in Europe. In a 1933 interview with Goering, Halton was struck by how deeply the Nazi leadership had itself drunk from the cup of its own delusions. What was worse than believing their ridiculous racial bigotries, especially against the Jews within and the *Untermensch* (literally 'sub-human') without, was their unfounded conviction that the policy makers in the West – especially in London – secretly supported them, even if they couldn't express this support openly. After all, it wasn't a secret that it was Western capital – sourced through London and Wall Street – that was financing the rebuilding of Germany, and which by necessity was close to the Nazi economic programme. The interests of both sides were therefore closely aligned. If the outcome of Versailles was to be reversed, Goering asserted, Britain and France would be sensible. Both countries knew that Germany was Europe's bulwark against the spectre of Soviet bolshevism. 'Germany will save Europe and Occidental civilisation,' he blathered. 'Germany will stop the rot. Germany will prevent the *Untergang des Abendlandes*.'*

Goering was well briefed about attitudes in the West. Many well-read people in London and New York were already saying, and believing, some of this although those in Warsaw, Prague and Paris were less inclined to do so. The unbelievers were widely ridiculed in their own countries. In Britain, Winston Churchill was the most prominent of those who did not accept this worldview, but for the most part the political establishment ignored and reviled him in equal measure. Churchill was, at the time, like a biblical prophet crying in the wilderness, scoffed at by those who considered him a warmonger for advocating robust responses to the militaristic posturing of the totalitarian states.

Yet all it needed, argued Halton, was for people – in Germany and the West – to read. The Twenty-Five Demands of the 1920 Declaration were unequivocal. They were the Articles of Faith of a new religion and held to fanatically by true believers. This was the foundation stone upon which all else was being built. Hitler mapped out these plans for rebuilding Germany in this image in *Mein Kampf*, but wisely refused his publishers permission to have the rambling tome translated into English. It would give the game away. When Halton interviewed Albert Einstein at a secret location on England's south coast in September 1933, he asked the famous scientist whether it was Hitler's plan to destroy European Jewry. 'Jewry?' Einstein retorted. 'Jewry has less to fear than Christendom. Can't you people read?' When Halton asked him what the ultimate result of the Nazi project would be, Einstein immediately responded, 'War. Can't the whole world see that Hitler will almost certainly drag it into war ...[?]'

It couldn't. Most Western diplomats and statesmen in 1930s Europe made the mistake of misunderstanding Hitler's true nature

---

*A reference to the book *The Decline of the West*, written by Oswald Spengler in 1926.

and ambitions. After all, these – articulated in *Mein Kampf*, in repeated public utterances in Germany, in speeches and interviews – were so fantastical, outrageous even, that most reasonable men and women did not consider them viable, and so dismissed them utterly. This was the underlying reality of appeasement: sane politicians and statesmen believed that Hitler likewise was rationally calculating what he could secure by beating the war drum without taking his country and people into a ruinous war. When they visited Hitler in person, the Führer mollified them with honeyed lips. He was a man, the representatives of the democracies believed, with whom they could do business. Did he not repeatedly assert that he wanted peace? The appeasers did not realise – until it was too late – that if Hitler could not secure peace on his own terms, he would do so by means of war.

On 21 May 1935, the American journalist William ('Bill') Shirer in Berlin recorded in his diary that Hitler had made yet another masterful speech in the Reichstag proclaiming his desire for peace. Yet, like Halton, Stowe and Thompson, among others, Shirer saw clearly through the noisy charade. Hitler was in fact calling for war, under the cloak of demands for concord. It was the assassin's knife, hidden until it was plunged, deep and red, into the body politic of Germany's enemies. Shirer was now convinced of Hitler's remarkable powers of oratory: what he lacked in visual presentation he made up for in his fantastical speeches. He held his audiences spellbound, but the demands he made that night in exchange for peace revealed that in fact they were impossible for Europe to accept. And what did he get for this drum-banging? Fear.

In Western capitals politicians were unwilling to stand by the provisions of the treaties jointly made at Versailles in 1919 and Locarno in 1925, primarily because there was no appetite to enforce them. After all, 'Germany was merely exerting her natural rights', the argument went, and 'Versailles was embarrassingly – and unnecessarily – harsh', a Nazi deceit still in currency today. 'We would do the same, surely, given similar circumstances. Give the Germans some leeway! They are no threat to us.' Using these arguments, Britain unilaterally allowed Germany at the Anglo-German Naval Agreement in London in June 1935 to break free of the Versailles straitjacket, providing it with the ability to build as many submarines as the British. 'Why the British have agreed to this is beyond me,' remarked Shirer, unpersuaded of the argument that embracing Germany would make it more peaceful. 'German submarines almost beat them in the last war and may in the next.'[4]

At a speech to the Reichstag on 8 March 1936, Hitler revealed this inherent – though carefully hidden – contrast between his demand for peace and his desire for war. After a diatribe against the threat of bolshevism, Hitler told Germany – for the first time – that he was unilaterally rescinding the Versailles Treaty with respect to the demilitarisation of the Rhineland. At the same time, he repudiated the 1925 Treaty of Locarno. This was all in pursuit of peace, he asserted, but one that was this time favourable to Germany and its interests. It mattered not whether these conflicted with those of its neighbours: to Hitler, only Germany mattered.

It was not just the carefully choreographed Reichstag that cheered Hitler. The repudiation of Versailles was endorsed by most Germans, even those who had no time for their country's leader. Rhinelanders, who certainly didn't want another war with France, as Shirer reported, had also caught 'the Nazi bug' and were hysterical about this supposed recovery of Germany's sovereignty, self-respect and self-determination.

Peace! It was that simple. Hitler prefaced all his military actions with the claim that his ultimate purpose was peace. He was right, of course, except that this wasn't what it seemed. Germany would accept peace *after* whatever war was necessary to reassert the rights due to it by its ancient birth right. Indeed, Hitler preached peace long and loudly. So much so that any accusation that he in fact wanted war sounded deliberately argumentative, even subversive. Yet *Mein Kampf* made it explicitly clear that Hitler saw war to be inevitable, and that peace was acceptable only on terms favourable to Hitler's concept of Germany's manifest destiny. Indeed, the one thing that the Nazi elite held to fanatically was the 25 demands of the Programme of the National Socialist German Workers' Party signed on 24 February 1920. The first four were:

We demand the union of all Germans to form a Great Germany ...

We demand equality of rights for the German People in its dealings with other nations, and abolition of the Peace Treaties of Versailles and St Germain.

We demand land and territory (colonies) for the nourishment of our people and for settling our superfluous population.

None but members of the nation may be citizens of the State. None but those of German blood, whatever their creed, may be members of the nation. No Jew, therefore, may be a member of the nation.[5]

The world did not understand the real Germany. Another journalist, Wallace Deuel, wrote in February 1941 that one 'of the chief reasons why much of the outside world has failed to understand the Nazis and what they have been doing and are planning to do is that people simply cannot believe that the Nazis are the kind of men they are'.[6] This truth lay at the heart of appeasement, and accounts in major part for its failure. Well-meaning visitors wanted to believe that Hitler and his henchmen were merely healing Germany by restoring its self-respect after the horrors of humiliation and the subsequent worldwide financial crisis. They were also imposing discipline following chaos. Surely no one, not even Hitler, wanted war? Was that not self-evident from Hitler's repeated speeches? It was ridiculous to assert that Hitler, now he was in power, would take Germany down the path of the extremist nationalism he had espoused in his immature political youth.

This view was superficial and naive, Shirer considered, because it saw the extent of Nazi politics only through the dangerous self-limiting prism of Western wishful thinking. Germany was, he believed, hard set on a path that would lead it to inevitable conflict with all – states, groups and individuals – who opposed him, and the evidence was everywhere. The ultimate end of Hitler's plan for Germany, and the purpose of all his policies – Four Year

Plans; 'guns before butter' speeches by the Nazi elite; and every evidence of the obvious resurgence of the Reichswehr (the army of 100,000 men that Germany was allowed to keep by the Treaty of Versailles) – was total war, Shirer thought: there was no other rational explanation. Despite the strictures of Versailles, none of which were being enforced by the international community that had imposed them, every public parade saw new and better guns, faster tanks, more aircraft and less butter. Matthew Halton agreed. At the end of 1933 he had written: 'During the last month in Germany I have studied the most fanatical, thoroughgoing and savage philosophy of war ever imposed on a nation. Unless I am deaf, dumb and blind, Germany is becoming a vast laboratory and breeding-ground for war ... They are sowing the wind.'

But few at home paid him any attention. It wasn't politically acceptable in the West to believe that war was the ultimate political ambition of Nazi philosophy, and of the new Nazi government. It was too preposterous for words. The failure of the democracies in the 1930s was the naive failure to accept that any sane, rational human being *wanted* war – sadly, much the same could be said of Vladimir Putin in February 2022.

Chapter 13

The slow rush to rearm

In retrospect, political events on the international stage from 1931 onward seem to have a horrible inevitability about them. The much-vaunted collective security of the League of Nations fell at the first hurdle when Mussolini invaded Abyssinia in October 1935. Neither the League of Nations nor any of the major powers were prepared to use their military capability to enforce the peace. The entire concept of collective security floated away into the ethereal mists of pacifism and wishful thinking. At the time for the British, the result was more commitments and stretch for their army. Britain sent reinforcements from all three services to the Mediterranean, the army despatching three additional battalions to Malta and a brigade to Egypt, followed thereafter by a battalion of light tanks (from the Tank Brigade), a company of medium tanks and a mechanised artillery brigade. From these units came the kernel of what in due course would be the 7th Armoured Division – the 'Desert Rats'.

The Rhineland then followed in 1936. The French Army, still large, did not act because there was no political will do so. France was politically fractured and unable to act decisively to protect the security status quo established by Versailles. Inaction was the loudest political message of the century, for by its silence both Britain and France told Hitler that he could act with impunity where German 'interests' were concerned. The advent of the

Spanish Civil War in July 1936, together with Germany's blatant involvement on the side of the fascist rebels under General Franco, was the entrée to the creation later that year of the 'axis' between Rome and Berlin and the subsequent Anti-Comintern Pact between Germany, Italy and Japan. In addition to the clear evidence of rapid German rearmament in 1937, Japanese armies launched attacks in China.

In the following year Britain's mandate in Palestine also began to bite, with a full-scale Arab rebellion against fast-rising Jewish immigration in 1938 further draining military manpower. Palestine was a critical component in Britain's security jigsaw in the Middle East, a fulcrum between Egypt to the West and Iraq and the Persian Gulf to the East. Instability was not welcomed at a time of heightened tension elsewhere in the Horn of Africa. A frightened Cabinet agreed a further enhancement to the air defence equation in February 1936, increasing the RAF target to 1,736 front-line aircraft to be in service by 1939. For Palestine the old model, in place since 1921, of the Royal Air Force providing security by means of a squadron of aircraft and two of armoured cars, required rethinking now that boots were needed on the ground. By 1938 two full infantry divisions – Major General Richard O'Connor's 7th Division and Bernard Montgomery's 8th Division – were attempting to put down the rebellion, curtail Jewish immigration and provide security against the depredations of any would-be enemy.

The period from 1933 through to the outset of the Second World War in September 1939 was one in which the army operated within a government policy that resolutely refused to consider the prospect of another military commitment to Europe of the kind the country had provided in 1914. Britain spent most of the two decades following the Great War assuming that it could absolve itself of responsibility for any further military engagement in Europe and made no pretence of preparing for it. It insisted that any future conflict in Europe was none of its business, with a message to its continental allies that was tantamount to: 'You can get on with your squabbles, but we aren't going to get involved. We are going to focus on protecting Britain from air attack, and we've got an empire to worry about. You're on your own.'

It was a reversion to the traditional British strategy of isolation. Britain would protect its national interest in Europe, it believed, by means of a 'limited liability' only, provided mainly by bombers of the Royal Air Force together with, and only when absolutely necessary, two divisions of infantry to protect Belgian airfields. It was, if the country was being honest with itself, in actuality a policy of 'no liability', overturned only following Hitler's advance on Prague in March 1939. This was the point at which a geopolitical reality – lacking for a decade and a half since Locarno – was finally injected into British decision-making. The policy of appeasement was reversed and funding for the rapid expansion of the army was unlocked.

For the army, however, it was too late. The increase that followed Chamberlain's *volte face* allowed solely for conscription and the manpower enlargement of the army, not its transformation into an effective warfighting organisation able to take on the Wehrmacht on equal terms. A lost two decades could not be recovered overnight. It was now too late to ask the crucial question: 'For what kind of war do we rearm?' The practical problems remained immense. The greatest challenge for the British Army in creating an expeditionary force foundered once more on the structural impediment of the Cardwell system. If an expeditionary force were constructed in Britain to service a new strategic requirement for the army – i.e. over and above that of an imperial gendarmerie – it would significantly affect the way in which the British Army's commitment to the defence of India was delivered. So long as the War Office considered the main threat to Britain's national interest to come from the East, that is where its defence priorities would remain, and defence inadequacies elsewhere would be accepted. Only when the shift to Europe occurred would the requirements of an army for the Continent be entertained.

Likewise, the idea that a modern mechanised combined-arms expeditionary force could simply be 'magicked' upon the troops available in Britain without a coherent fighting doctrine, training and preparation and with the only troops available those assigned by Cardwell, was abject foolishness. To undertake a change of this magnitude required political and military vision and leadership in both Britain and India to reconfigure Cardwell to allow for the creation of a modern, deployable expeditionary force. Neither was forthcoming. As with much other military doctrine in this period, it was a lack of clarity about *how* the army would need to fight that led to a failure to provide direction for the future. Notwithstanding the written doctrine that did exist, successive CIGS, from Milne to Montgomery-Massingberd and Deverell,* were not convinced of the need to lead an army debate about the type of doctrine required for future war. That which existed – the Field Service Regulations 1935 and Field Service Regulations Volume 1 Organization and Administration 1930 – were insufficient for the task. They were good documents insofar as they went, but they could not reflect the experience of combined-arms operations at brigade, divisional and corps level because the British Army had not developed the techniques required to convert these ideas into practice. The doctrine that existed did so in a vacuum to the experience of 1918; nor was it part of the DNA of the army. The last formation exercises had taken place in 1934. It was thus hardly surprising that successive Secretaries of State for War were unable to provide the necessary direction for a type of warfighting they couldn't themselves articulate because the army could not describe it to itself. Duff Cooper, who took the appointment in November 1935, urged a reform of Cardwell, but was unable to do much more than recognise that the problem required resolution. A root and branch study of the British military system was put off until 1939. This was in part because of the distractions in government caused by the intensity of the arguments over the continental commitment.

---

*Field Marshal Sir Cyril Deverell became CIGS on 15 May 1936.

Cooper, who was more knowledgeable of military matters than any other Secretary of State during this period, nevertheless found himself engaged in a battle with Chamberlain – one that he was always going to lose – on the subject of preparing and equipping an army for war in Europe. Chamberlain was an untiring exponent of limited liability and Cooper a trenchant supporter of the need for a continental military commitment. Chamberlain simply did not see the need for a large army, and certainly not one for deployment in Europe. The Cabinet debated the subject between December 1936 and May 1937, during which the political relationship between the two men broke down. But for all his defence of the army's prerogatives, especially of the need properly to equip the Territorial Army, Cooper was able to do nothing to reform the army's fighting doctrine. There is a telling phrase in his memoirs which hints at this, where he admits that he 'acquired little credit during my tenure in the War Office'.[1] Given his spirited arguments for preparing for the war he now believed to be inevitable, he wasn't commenting on these exertions, but on a deeper failure to comprehend the nature of the fight that was coming, and urgently to prepare a dispirited, confused and directionless army for what it was about to confront. It was in this failure of imagination and vision – to recognise how the army needed to change to meet the worsening international situation, even to the extent of changing or scrapping the Cardwell system – that was to lead Britain to disaster in 1940.

During the rearmament debates in the Defence Requirements Committee and the Cabinet over the years since 1933 the British Army was hamstrung because it had no clear view as to the type of modern army it needed. Most of the conversations about rearmament that took place were concerned about the numbers of units available to meet all of Britain's global commitments, not about *how* it would have to fight. If the British Army had had a *warfighting* doctrine, gifted and paid for by the Cabinet, requiring a comprehensive and integrated fleet of armoured vehicles (tanks, infantry carriers and mobile artillery) together with an integrated tactical air force, the discussions about priorities would have been very different. But there were no such discussions.

As a result, Britain, when it mobilised, ended up in September 1939 with a 1914 army carried in lorries, many of which were civilian vans impressed into military service. By the start of the Second World War the army had only a light tank in its order of battle: a decade of indecisiveness and lack of funds had not even replaced the old medium tank of the 1920s Experimental Mechanized Force. The lack of clarity – for want of an effective *warfighting* doctrine – about 'cruiser' tanks (which could travel at their own speed) or 'infantry' tanks (which travelled at the speed of marching infantry) had still not been resolved. Nor had a decision been made as to whether the army needed light tanks (for the empire) or heavier ones for Europe. Confusion reigned. Until 1938 Britain's single Armoured Division – comprising a mixture of light tanks and armoured cars – was designated in any case for Egypt (two other armoured divisions were in the process of being formed at home, slowly). As a result, when rearmament happened it served to fill up the equipment and manpower gaps in the five divisions – four infantry and one cavalry – of what was to become Britain's deployable force (the BEF), all of which were units diverted from the Cardwell allocation.

Parliament had quibbled with the initial Defence Requirements Committee recommendations in 1934 and the resulting compromise allowed little for modernising the army and in any case was quickly found to be inadequate. By 1935 it was clear that German rearmament was marching ahead with remarkable pace. In addition to the dramatic growth of the army, the Luftwaffe had already exceeded the size of the Royal Air Force and would soon achieve parity with France, which had 1,500 aircraft. The Cabinet instructed the Defence Requirements Committee to revise its figures for aircraft production, and a new plan was agreed in July 1935, in which 1,512 aircraft would be built and be in service by 1937. A White Paper published on 3 March 1936, *before* the reoccupation by Germany of the Rhineland in 1936 and owing much to Chamberlain's persistence, committed to a massive programme of rebuilding Britain's armed forces.[2]

The line taken by the Chiefs of Staff hadn't changed since their first report, but the situation was rapidly bearing out their repeated warnings. They argued consistently that the country needed to take seriously the threat of a war on two fronts, with Japan in the Far East and Germany in the West. The time had now passed simply for deficiencies to be made up, as they had recommended three years before, but for a massive programme of rearmament. So far as the Cinderella at the ball was concerned, the Chiefs of Staff made a detailed case for the need for the army to provide a BEF of five regular divisions and a tactical air force for despatch to the Low Countries, followed up by reinforcement of four Territorial divisions in four months, a further four Territorial divisions in six months and a final two divisions two months after that. Duff Cooper, the Secretary of State for War, did his utmost to persuade his colleagues in Baldwin's government to support these relatively modest proposals.

The arguments still failed to persuade the Chamberlain-led Treasury and Cabinet duly discounted the proposals. However, an additional four regular battalions were to be added and the Territorial Army modernised and at long last the commitment to a five-division BEF was accepted, although it was carefully described as a mobile force available for despatch to quench prospective fires anywhere in the empire, not just in Europe. Making a commitment to the despatch of Territorial divisions to Europe would have too great an impact on the economy, and it also raised the prospect – unacceptable at the time – of conscription.

The March 1936 White Paper proposed to increase the number of cruisers in the Royal Navy from 51 to 70, with two new battleships and an aircraft carrier laid down. The relentless rise of numbers in the Royal Air Force would also continue, increasing to a target of 1,750. In terms of the country as a whole, the rearmament jump was significant, as the figures for defence expenditure as a whole demonstrate:

Year	£ millions	Percentage of budget	Increase over 1932–33 baseline
1932–33	103	12.9	0%
1933–34	107.9	14.9	4.54%
1934–35	114	15.5	9.65%
1935–36	137	17.7	24.8%
1936–37	186.7	22.4	44.83%

The respective percentage of Gross National Product between Britain and Germany was:

Year	Britain	Germany
1932	3%	1%
1933	3%	3%
1934	3%	6%
1935	3%	8%
1936	4%	13%
1937	6%	13%
1938	7%	17%
1939	18%	23%
1940	46%	38%

Even though rearmament began as a policy in 1935 and 1936, it did so fitfully, and in the face of extreme reluctance from the Treasury, and some political opinion which held that the three services were agitating to make the most of a deteriorating international situation to feather the nests of their own service.

This half-decade was in fact a period of remarkably harmonious union among the three service chiefs within the Chiefs of Staff Committee (under its chairman Lord Chatfield), as they strove to repair the damage made by a decade and a half of decline in Britain's defences. But there was no sudden Damascene moment across government about the parlous state of Britain's defences or of the threats posed to Britain's national interest by Germany,

Italy and Japan. All the way up to the moment that Hitler marched into Prague in March 1939 there were very powerful and influential voices in Whitehall urging caution in respect of the financial demands of rearmament. For while the rebuilding of defence capability following the establishment of the Defence Requirements Committee in 1933 was accepted, the nature and scale of rearmament remained an issue of debate and at every step there was opposition by the Treasury to its extent in the context of the size of Britain's purse, notwithstanding its reluctant acceptance of the principle.

This isn't to suggest that the Treasury wasn't right to protect the public purse from profligacy. But for this half-decade the Treasury was out of step with the growing evidence of a slide to war. It insisted throughout that, despite the increased spending required, it should be done within a framework of 'business as usual' for the economy. There was to be no change in industrial policy, or priority given to defence contracts with industry, or any diversion of skilled labour from its existing tasks. There was to be no building of factories solely for military needs, nor any re-tooling of machinery for the building of arms where it might be needed for income-earning production. The Treasury's hope was that before long diplomacy or appeasement would solve the security problems in Europe, and defence spending could then be as quickly switched off, with no negative impact on the direction of the economy. In any case, the betting in government was still on Herr Hitler not having any evil intentions with regard to the sovereignty of Germany's neighbours.

The story of the process of rearmament was principally one about the growth in air power. The popular consensus that the Royal Air Force would be the arm that protected the British Isles from attack allowed it to soak up much of this cash, not just for the development of inadequate bombers, but fortunately also, in the mid-1930s, for the development of excellent fighters. The story of the ineptitude of the former programme, and the vast sums of money disastrously wasted for no or limited effect, is told elsewhere.[3] Decisions made by the Cabinet in this period, until 1939, failed to acknowledge that war in Europe would almost certainly require a land forces response. Likewise, military power could also have had a deterrent effect on the European adventurers, although this self-evident truth was not considered as a matter of policy. The concept of military (i.e. army) deterrence is a subject remarkably absent from the writing and debates of the time.

Equally, the period was one in which the defence community, represented by the three service Chiefs of Staff, had considerably less authority than the Treasury did in the Cabinet. The political leadership of the country also consistently ignored the military advice it was given. The concept of limited liability in Europe was the product both of a revulsion at the prospect of another land war like the last and of a profound unwillingness to put British boots once again on continental ground. The fundamental failure of Chamberlain's rearmament programme was that it refused resolutely to acknowledge that the army would be required once more to fight a substantial war in Europe. Equally, the failure of the War Office and the army as a whole was to prepare – if only intellectually, mentally and doctrinally (given that modern equipment: armoured artillery, armoured infantry carriers, tanks and so on simply weren't available in Britain's armoury at the time) – for a new and substantial *warfighting* task.

The principal political cheerleader of limited liability, and the most responsible for consistently ignoring the advice of the Chiefs of Staff – not just and not only for reasons of money, but because he considered himself a superior strategist – was Neville Chamberlain. The man who became prime minister in March 1937 is often lauded for his commitment to rearmament, which is used to expiate the sins of appeasement. But it was his repeated refusal to consider the creation and deployment of a warfighting force to Europe that was the most cataclysmic failure of this period of Britain's foreign policy. The Second World War was to demonstrate that the opposition to a continental commitment was strategically, geopolitically and technically naive. It was strategically naive because it assumed that both the Royal Navy and the Royal Air Force would be sufficient to protect Britain's interests in Europe. It was geopolitically naive because it failed to recognise the criticality of a military alliance with France, if Belgium – vital to the defence of the English Channel and to prevent Belgian and French airfields being used to attack England – was to be protected. And it was technically naive because of the grand assumptions bandied about by the air-power theorists about the overwhelming importance of the bomber in a future war. These assumptions failed to acknowledge the need to fight on land, and to prepare to do so. By his determination to protect the army from future bloodshed and his insistence on limited liability, Chamberlain in fact denuded it of the capabilities it would need when the inevitable happened and a highly intensive war came once again to Europe.

In contrast, from 1933 the new regime in Germany did all it could to *imagine* future war. It couldn't prepare for it in *actuality*, because of the strictures of Versailles that limited its army to 100,000. But this didn't stop the Reichswehr between 1919 and 1926, under the leadership of General Hans von Seeckt, from developing new theories and doctrine, or from undertaking extensive exercises (with and without troops) and wargames. Within its 100,000 manpower cap was a disproportionate number of officers, the backbone of any future expansion. The Germans also built on the ideas of the British theorists.* Whatever the truth of the matter, Liddell Hart's approach to the idea of the armoured division – a mixed force of armour, armoured infantry, armoured artillery and tactical air power – designed to break through enemy depth positions into an expanding torrent that would unravel enemy rear areas and

---

*A point seemingly confirmed by General Heinz Guderian in his memoirs: Heinz Guderian, *Panzer Leader* (London: Michael Joseph, 1952), p. 20.

discombobulate their command and control, was essentially the same as that adopted in Germany, but not in Britain.

In Germany, the first three armoured ('panzer') divisions were formed in 1935. The panzer division (a mix of tanks and lorried infantry)* was designed to be the lance or spear punching through and between an enemy's fixed defences (the 'infiltration' of Ludendorff's 1918 Spring Offensive tactics), be they presented in linear form or by means of bastions scattered across a deep battlefield. The purpose of the lead units in an offensive – the panzer divisions – was not to stay and fight, but by means of speed and manoeuvre to drive deep into the heart of the enemy, there to so confound the enemy that they would become confused, disorientated and disheartened. They would leave the enemy positions they had passed to be mopped up by forces travelling up behind. Follow-up infantry, walking or horse drawn, could lay siege to whatever enemy bastions remained, cut off from their lines of communication and relatively easy therefore to reduce.

———

Neville Chamberlain moved from being Chancellor of the Exchequer in Baldwin's government to prime minister in March 1937. The most powerful man in government became the most powerful man in the country, and the empire. He appointed the reforming Leslie Hore-Belisha as Secretary of State for War to replace Duff Cooper, who had so irritated him for his support of a continental commitment.† Hore-Belisha in turn brought with him Basil Liddell Hart as his military adviser (the marriage was in fact arranged by Duff Cooper). One of their first jobs was to sack the

---

*Each 1939 panzer division had 400 tanks and a motorised infantry brigade, together with reconnaissance, engineer, artillery and anti-tank troops. In 1942 each panzer division was reduced to 200 tanks and two lorried infantry brigades.

†Cooper was moved to First Lord of the Admiralty, but became an increasingly strident opponent of Chamberlain's policy of appeasement. Chamberlain, in return, saw his colleague as a warmonger.

CIGS, General Sir Cyril Deverell – by leaving a note dismissing him on his desk – after a period in office of only 18 months. This was based on advice proffered by Liddell Hart, arguing that Deverell was too conservative and an impediment to reform. That may well have been true (the charge against him was undoubtedly exaggerated), but the manner of his removal, by a Secretary of State advised by a contrarian newspaperman, was destined to set the senior echelons of the army firmly against both Hore-Belisha and Liddell Hart.

The man chosen to replace Deverell was Lord Gort VC. A fighting soldier, Gort was to find himself seriously out of his comfort zone in the gladiatorial pit that was Whitehall, a place where, in any case, the army and its interests still sat firmly at the bottom of the political pecking order. Hore-Belisha's ambition was to modernise the army. To his credit, much was done to improve the army in a wide range of areas, including mechanisation. In 1935, when the process of mechanisation began, it was decided to convert the cavalry to engines, rather than to build an expanding tank arm, a decision made not on the needs of doctrine (i.e. 'creating an armoured division because of the need to manoeuvre operationally') but on expediency (i.e. cavalry doctrine had not changed, but technology had determined that the engine would be more helpful on the modern battlefield than the horse).

The Mobile Division that emerged seemed to have been created on the basis of what automotive power money could buy, rather than what the science of war determined was required. It would comprise a reconnaissance element of two armoured car regiments, two mechanised cavalry brigades, a tank brigade (with light tanks) and two artillery brigades with tractor-drawn guns. Under Liddell Hart's influence (as adviser to Hore-Belisha) this composition was changed in 1937, the Mechanized Division for the first time having a far less 'mechanised horse' feel about it. The cavalry were all converted to light tank regiments, and each brigade was accompanied by a lorry-borne infantry battalion. Slow steps were being made towards a modern manoeuvre formation, but it was progress made not on the basis of agreed doctrine, but on that of predilection, influence and special pleading. It was no way to design an army for war.

However, the really important task which should have been pre-eminent above all others – to determine what the army was for in this highly combustible age, who it was likely to fight and how it would have to fight – remained unexamined. In addition, with Liddell Hart on Hore-Belisha's arm, any commitment to a future European war remained anathema. Chamberlain and his supporters genuinely believed that there would not be a land war in Europe, or at least not one involving Britain.

Chamberlain's primary concern remained the impact of rearmament on the economy. His view that the process of rebuilding Britain's defence capability should focus on the Royal Air Force at the expense of the army never altered. The Chiefs of Staff estimate for the cost of the latest revision to the Defence Requirements Committee Report in October 1937 was £1,605 million, which far exceeded the £1,500 million the Treasury had agreed was available. But even that figure of £1,605 million was quickly overtaken by events, with the decision made to build yet more aircraft, raising the total number of airframes by 25 per cent over that agreed only the previous year. Rearmament placed enormous pressure on the industrial base of the country and required a diversion of scarce materials and skilled labour. But there was still no acceptance that the rapidly increasing threat from German rearmament might also affect the army. The priorities for the army in late 1937 remained the defence of the United Kingdom as the hub of the empire; the protection of its sea lanes of communication; the defence of its overseas territories; and, only when these had been achieved, support to Britain's allies in the near abroad. The army was instructed to consider the despatch of an expeditionary force to Egypt, rather than to France. At home, it was suddenly lumbered with another undesired commitment: anti-aircraft defence, to shoot down the legions of bombers confidently predicted by the air power theorists.

Chamberlain regarded these moves to be a prudent, if expensive, insurance premium. The only alternative to planning for war, he considered, was to persuade Germany not to follow the war path. Most ministers and the Chiefs of Staff agreed with him but remained frustrated that the seriousness of preparation for war was still being hampered by economic considerations, when all the evidence showed that the risk of war was increasing. No further evidence of Hitler's ambitions was needed once the Austrian *Anschluss* ('annexation') took place in March 1938. Chamberlain was forced to announce to the House of Commons on 24 March of that year:

> In order to bring about the progress which we feel to be necessary, men and materials will be required, and rearmament work must have first priority in the nation's effort. The full and rapid equipment of the nation for self-defence must be its primary aim.

This did not mean, however, that the army was given a bone. The emphasis on rearmament remained on the Royal Air Force, with a further increase made in front-line aircraft, mainly fighters, and the introduction of radar. The idea of a continental commitment remained anathema to both public and government, and France was told not to assume that any troops would be available to support it if Germany invaded. But at long last the principle of 'business as usual' ended, and low-level staff talks with the French Army were initiated in May 1938. France knew precisely where Britain's security priorities lay and did not take these talks at all seriously. What was on offer was no more than two divisions within 16 days of the start of war.

The gradual change to the attitude to a continental commitment came about as some of the earlier security certainties began to crumble. First, the question was asked as to whether it was possible to deter Germany from war against France if Britain had no publicised plan to reinforce its allies in Europe in the event of war? It was thought possible that, given France's political fragility at the time, without Britain's direct military support on land it might cave in to German demands and secure an accommodation that would not be in Britain's interests. Secondly, the role of the Royal Navy in replicating the Great War strategy of blockade was looking far less credible with Hitler now able to secure food supplies from elsewhere in Europe. Thirdly, the old arguments about the Royal Air Force being able to strike at economic targets in Germany and bring the country to its knees were now looking frankly foolish. Neither the numbers of bombers nor the range nor the payload of any RAF bomber in service or expected in service in the near-future could hope to do anything near what the theorists had for so long preached. It was hardly surprising that Trenchard's bomber strategy was quickly superseded by one in which, under Air Chief Marshal Sir Hugh Dowding, commander-in-chief of Fighter Command, Royal Air Force, from its formation in 1936 until November 1940, the enemy would now be defeated by swarms of nimble fighters. The one element of the air power equation that was only to receive the crumbs from the table was tactical air support to land formations. It wasn't until 1944 that this situation was to be satisfactorily remedied.

For the army, 1938 remained in retrospect a frustrating and wasted year. Although the War Office had built a plan to create and despatch an expeditionary force of four regular infantry divisions and two mobile divisions to France and a further two divisions to the Middle East, and to equip and train the first four Territorial divisions, opposition from the Royal Air Force and Royal Navy to the idea of a commitment to Europe remained, as did the absolute veto of Cabinet. United though the Chiefs of Staff had been during the 1930s the fundamental weakness of Britain's position was that the country did not have a coherent security strategy that balanced the air, land and sea needs for the defence of the home islands. The army remained concerned with its role of defending the empire. Only a tiny few had given any thought to the problem of how to fight a peer-on-peer war on the Continent. The navy focused on its battleship fleet and the threat in the Pacific from Japan. The air force was taken up with the idea of the bomber offensive as a strategic deterrent. No one expended any energy on working together to create a single – combined-service – view of Britain's defence effort. The Chiefs of Staff Committee provided a committee's view of three separate and competing sets of defence priorities, not a single view of a single solution to a complex problem with three competing but equally valid angles. This would require a Ministry of Defence, but even that would not see the light of day for many years after the end of the Second World War.

It was left to Gort, the new CIGS, to keep on plugging away at the argument for an expeditionary force. He argued consistently that in the event of catastrophic war in Europe, Britain would have no choice but to intervene, and to do so on a substantial basis. Pretending it wouldn't happen, or that war would force the French to extend the Maginot Line to the Channel, would not make the war go away. Assuming that Britain could stay out of a future European imbroglio was as unreal as it was naive. It wasn't warmongering to prepare for this eventuality, but the very prudence that Chamberlain preached. In the event it wasn't until February 1939, days before Germany invaded the remainder of Czechoslovakia, that Chamberlain reluctantly accepted Gort's plans.

———

After the conclusion of the Munich Agreement in September 1938 Hore-Belisha had a sudden and Damascene conversion to the need to prepare the army for a deployment in Europe. Despite his earlier acceptance of the limited liability argument, he now saw that a military commitment was inevitable. Politics and ideology aside, it was now time to prepare the army for such an eventuality, and he started making himself a nuisance in Cabinet on this subject. In February 1939 agreement was reached to finance a 'field force' (the new politically correct name for the Expeditionary Force, a title with unacceptably 1914 overtones), although initially this extended only to four regular divisions and four Territorial.

It took the seizure by Hitler of Prague in March 1939 to change Chamberlain's mind but when it happened this was equally dramatic. Appeasement was now dead. All the years of ideological dissimulation were overturned in a moment. Without bothering to discuss this with the Chiefs of Staff, a hallmark of his previous disdain for the advice from his service chiefs, Chamberlain doubled the size of the Territorial Army to 26 divisions. The army was to have 32 divisions, six regular and 26 Territorial. The commitments Britain had felt unable to give the previous year to Czechoslovakia were now handed out with abandon to Romania, Poland and Greece. Bizarrely, Poland was told that it could expect no RAF bomber attacks on Germany for fear of retaliation on Britain. On 20 April conscription, the one thing that would denote a commitment to service on the Continent over anything else, was introduced before war had even been declared. It had been the fear of another mass army of young men secured through conscription that framed society's reluctance to give its support to the army after the Great War.

The immediate consequence for the army was that it wouldn't have the chance to build up a small, well-equipped and capable mechanised force as a test bed for further development, able to deploy to Europe and fight any modern enemy. Instead, the rapid expansion of the army meant that a new citizen army would now be created, on an unsophisticated infantry basis of men and bayonets that had mirrored 1914, capable of digging in and fighting a war like that of the early stages of the one that had just passed. In March 1939 Hore-Belisha raised the establishment of the Territorial Army to an extraordinary 340,000 men. Then, in April, he persuaded Chamberlain to agree to conscription. Within weeks 200,000 men arrived to begin training, stretching the regular army to breaking point as it attempted to train, equip and absorb the newcomers.

The problem with expanding for war was that the army grew to grow, rather than growing for a purpose. The army knew that it needed numbers of men in uniform – at the time the Germans were mobilising 105 field divisions, with six panzer divisions and a further ten motorised divisions – but did not know what *warfighting* structure to place them in, other than expanding that which existed. The two decades after 1918 had been wasted in a directionless muddle. In 1939 the British Army had a half-baked conception of warfare, a mix of the pre-1918 linear concepts mixed with dollops of armoured warfare theory contained in the 1933 Field Service Regulations (written by Fuller). Some tanks would fight on their own, in tank brigades (although the 'cruiser' brigades didn't yet exist) while the infantry would fight as they always had, supported by heavily armed Leviathans, which also didn't exist in the numbers required. As Corelli Barnett observes with characteristic acerbity:

> Part of this critical British failure in the Second World War was owing to the lateness of the British programme of modernization; part to indecision and dithering over specifications; part to the sheer technical incompetence of some British engineering firms; part to pre-war Treasury meanness.[4]

The soul-destroying effect on the army during this period is perfectly summarised by the historian of the 3rd Carabiniers:

> Since the war a large number of military theorists had appeared in the United Kingdom, basing their conclusions on the type of fighting carried out on the Western Front, and endeavouring to prove that such slaughter of civilians in uniform could be avoided in the future by the establishment of a small professional army making up for lack of numbers by the extreme mobility made possible by mechanical vehicles. 'God', it was said, 'was no longer on the side of the big battalions.' Although this comfortable assumption could not stand up to logical argument, and the idea that the pace of an army in the field could be speeded up from the three miles an hour of the infantryman to the forty miles an hour of the motor car was completely false, their theories nevertheless had a considerable influence on military thinking. It was however impossible for soldiers to do more than theorise,

for expenditure on the Army had been cut to such an extent that deficiencies in all establishments, men, arms, vehicles, and equipment, put a stop to any form of realistic training.[5]

The army was constructed in accordance with what the government believed the country could afford, not on an analysis of what was required. Worse even than the procrastination, technical incompetence and financial short-sightedness that meant that on 3 September 1939 the British Army had only received *two* of the new Mark II infantry tanks, the 'Matilda', from the total Mark I and Mark II tanks that had been ordered was the fact that, in contrast to the position in the German Army, there was no clear view about how to fight. Britain had squandered the opportunity to use the limited force it possessed in 1936 to resist the Rhineland reoccupation, and thus end the possibility of future German aggrandisement in Europe. Three years later it went to war scarcely more capable militarily, swept up in the inevitable confusions and disorganisation of a general mobilisation for a war which it had denied for 20 years that it would ever need to prepare for, or fight.

The rapid march to war in 1939 made General Hastings Ismay, now chairman of the military element of the War Cabinet secretariat, angry. 'I was not frightened or even excited' at the onset of war, he recalled:

> but I was furious – furious with ourselves as with the Nazis. Less than twenty-one years had passed since the Germans had lain prostrate at our feet. Now they were at our throats. How had we been so craven or careless to allow this to happen? The Cenotaph was almost on the doorstep of our office, and every time I passed it I felt a sense of guilt that we who had survived the First World War had broken faith with those who had died ... They had given their lives in the belief that they were fighting the war which was to end all wars. And now their sons and grandsons were about to be sent to the slaughter.[6]

Britain had won the war, but had lost the peace.

If Britain and France had had well-trained, well-equipped and well-rehearsed forces on a par with what the BEF had possessed in late 1918 when the Great War ended, the Second World War might have started and ended very differently. It might not have happened at all. Action against Germany at the point at which Hitler remilitarised the Rhineland in 1936 would undoubtedly have warned Hitler that the Western Allies were determined to preserve the sanctity of the Versailles Treaty, and thus to resist an aggressor state unravelling the new sovereignties established at the end of the Great War. Versailles was concerned not merely with the issue of German reparations, but with protecting the many new borders established in the re-mapping of Europe. Then, in 1939, when Britain and France declared war on Germany, the Wehrmacht was looking east, into Poland. The Siegfried Line protecting western Germany was guarded by some 43 low-quality divisions. A confident Anglo-French army, built on an offensive doctrine inherited from 1918 and equipped and trained to those standards, could have unravelled German efforts at future territorial aggrandisement with relative ease, had it so desired. As history tells us, neither Britain nor France had such desire. 'One might have expected the French Army, on which such high hopes were placed, to have drawn off some of the weight of the *blitzkrieg* on Poland by an attack from the west,' observed Ismay. He noted that French arguments against undertaking an offensive were subsequently shown to be completely flawed. Britain, 'without even a platoon as yet on the Continent' was in no position to argue the matter.[7]

Chapter 14

Feeding the crocodile

By the beginning of 1938 Hitler had become greatly emboldened. His repeated demands for the so-called racial 'self-determination' of the estimated 10 million German speakers who lived outside the borders of the Reich, even though most of these had never been part of a united German state – the very first demand of the 1920 Nazi Party Declaration – became daily more strident. The reaction by Britain and France to both his rhetoric and his actions during the previous 18 months had been derisory. Many politicians – of every ilk – believed sincerely that to oppose Hitler would be to enrage him and bring about the very situation they were trying to avert. In March 1938 Paris and Moscow offered Czechoslovakia the promise of military support if it was threatened. London refused to make any similar commitment. As it turned out, France's pledge proved to be entirely hollow. Most politicians in France and Britain did not believe that the Nazis' radical agenda was real, considering that it was, rather, a mere ploy for domestic power. Once that power had been attained, like any other rational man, Hitler would follow a more statesmanlike course, especially where his neighbours were concerned. Provoking Nazi anger would, so they believed, only encourage the fanatics who believed this hegemonic nonsense about German racial superiority.

In any case, Nazism wasn't bolshevism, and the greatest fear of property-owning democracies in Europe since 1917 had been that of communist revolution in their own countries seeping out of the unrestrained violence that had overwhelmed the old Russia. Hitler's professed hatred of and opposition to bolshevism was enough for them. Little did they realise that Nazism and bolshevism were not opposites but rivals for the repression of liberty, competing for who would wear the jackboots. Accordingly, none had lifted a finger to prevent the remilitarisation of the Rhineland in 1936, in contravention of Versailles, despite the German General Staff expecting a fierce French response. Instead, France watched supinely, deep fractures in its political system preventing it acting with unity and authority. Britain saw no need to intervene, the consequences being judged far greater than any benefits achieved in forcing Hitler to abide by Versailles, though Anthony Eden observed presciently that the event would mean war in Europe within two years.[1]

France had in fact weeks before asked Britain for a military alliance to prevent precisely what was soon to take place. Britain's military weakness was at least one of the reasons for the British government's lassitude. The absence of an alliance with Britain led to France reluctantly refraining from ordering a military counter against Germany. It was the missed opportunity of the century. Hitler had retained a residual concern for the British Army he remembered from Flanders and might have withdrawn his troops at the first strong whiff of gunpowder. To his surprise, and relief, he encountered merely somnolence and timidity. The Wehrmacht stayed, the German people rejoiced, Hitler was emboldened, and war was guaranteed.

The year before the remilitarisation of the Rhineland, the United States had enacted the so-called Neutrality Acts, legalising America's descent into isolation from political entanglements abroad – specifically Europe – by making it difficult to trade with any warring country regardless of the righteousness of the cause. The result was that European democracies could expect no help from the United States if they were threatened by aggressors. Hitler judged – correctly, as it transpired – that if the European democracies were unwilling to back even their own treaties (Versailles and Locarno), they would be unwilling to withstand demands that had no direct consequences for their own sovereignty or security. For those countries with significant Germanic populations, such as Austria and Czechoslovakia, or indeed any country that did not cede to Germany's demands, the implication was clear. Complete and unequivocal acquiescence to Hitler's requirements was his term for peace in Europe. If he could not secure what he wanted, war would be the result. In the logic of the governing Nazi elite, the resulting war would be the fault of those who failed to acquiesce.

Fearful of provoking the beast, free Europe did nothing to stop the *Anschluss* with Austria in March 1938. Chamberlain admitted to the House of Commons during the crisis the reality that Britain could have done nothing to arrest 'this action by Germany unless we and others had been prepared to use force to prevent it'.[2] Their failure to act was the writing on the wall for Czechoslovakia. In London the journalist Virginia Cowles reported that the *Anschluss* raised tension 'to a higher pitch than at any time since the Great War'. Silent crowds gathered in Downing Street to watch the cabinet ministers leaving their hastily summoned meeting, while newsboys cried out to a cold, grey world: 'Germany on the march again.'

> Worried speculations ran the gamut from saloons to fashionable London drawing-rooms. There was a general rush of volunteers to ambulance services and air-raid precaution organizations, and hundreds of young business men signed up with the Territorial Army. Everywhere there was a cry for more arms.[3]

When Virginia Cowles had first arrived in London in 1937, an article she had written on the situation in Spain for *The Times* – which was editorially in support of the principle of appeasement if it served to prevent Britain from going back to war – caught the eye of Sir Robert Vansittart. One of the loudest anti-appeasement voices in the government, his admonitions made him one of a few in London arguing against negotiating a compromise with Hitler. He agreed with Cowles's argument that allowing Germany and

Italy to wage war unchecked in Spain was tantamount to letting the bully terrorise the playground. Indeed, he considered British neutrality on Spain to be craven, because Britain chose not to police the Non-Intervention Agreement it had signed with Italy, Germany, Russia and other European countries in August 1936, while Germany, Italy and Russia, in the full lens of world publicity, blatantly ignored it. Germany and Italy were 'trying to lead the world back to the Dark Ages,' Vansittart told Cowles, 'and if we don't wake up in time, here in England, they may succeed in doing it.' Cowles recognised that Vansittart's – as well as Churchill's – voice was being drowned by the noise of those who believed that peace needed to be given a chance, and at any price.

In London the government tried to assuage people's fear by declaring that tensions in Europe remained just that – issues for Europeans, and not for the British. Otherwise intelligent, rational commentators bent over backwards to demonstrate that everything was not as it seemed. Cowles noted:

> The *Times* ran leading articles emphasising the enthusiasm with which thousands of Viennese welcomed the Nazi regime and the Archbishop of Canterbury rose in the House of Lords to say that Hitler should be thanked for preserving Austria from a civil war. 'Why all the gloom?' cried Lord Beaverbrook, and the catchword stuck. 'Why all the gloom?' echoed the public and settled back in its comfortable illusion of peace.

In Berlin, an astonished Shirer watched the world – which was failing comprehensively to understand these ideological certainties – attempt to placate Germany. The thing people feared most was a return to the horrors of the Great War, so in their panic they allowed themselves to ignore their democratic principles, together with their international commitments – including their promises to protect Czechoslovakia – in the headlong flight to avert a war. Ironically, war was made more certain by their failure to demonstrate to Hitler the absurdity of using force, a folly first demonstrated in 1936. Under inexorable pressure from

Berlin, the only option seemed to be to encourage this new-age bully's victim – Czechoslovakia – to give in to its persecutor. If, by making territorial concessions, Czechoslovakia could prevent central Europe from falling into war, surely that would be a good thing, for which the entire world would be grateful? The argument fell, Shirer believed, on many points, not least because of the relentless militarism of the German state. Every day evidence appeared on Germany's streets of a march to war that would not be removed by the forced kowtowing of states being bullied by Berlin. But even Shirer was beguiled by the rationality of his own logical mind. In Geneva on 9 September 1938, watching the ineffective and deeply divided League of Nations do nothing about the looming Czech crisis, he dismissed the idea that Germany would fight to recover the Sudetenland, first because 'the German army is not ready; secondly, the people are dead against war'. Neither, however, were obstacles to Hitler. Berlin pumped out grievous propaganda that daily traduced, belittled and mocked Czechoslovakia. Hermann Goering shouted during the Nazi Party's annual rally at Nuremberg: 'A petty segment of Europe is harassing human beings ... This miserable pygmy race [the Czechoslovakians] without culture – no one knows where it came from – is oppressing a cultured people and behind it is Moscow and the eternal mask of the Jew devil.' As Vladimir Putin has shown in the third decade of the 21st century, peddling lies for political and military purposes did not die with Hermann Goering in the fifth decade of the 20th century.

The very real fear of war wafted across Europe in September 1938 like a cloud of poison gas. In Britain, egged on by a decade or more of catastrophising by the air power advocates, newspapers predicted millions of deaths from bombardment from the air. On 12 September, Hitler, in a speech to the party faithful in Nuremberg, raged at the Czechoslovaks, demanding that the slavery the 'German' population had suffered at the hands of their culturally impoverished overlords be ended by means of ethnic Germans being able to secure their self-determination. By this he meant, of course, that these regions should be transferred to Germany, regardless of the principles of national sovereignty or international law. His demand was the 'return' of all parts of Czechoslovakia where at least 51 per cent of the population considered itself 'German'.

The danger that Hitler faced was moving too quickly, and by so doing provoking war with France and Britain. Instead, to get his way, he turned up the notch gradually. He wanted acquiescence not by force but by a style of diplomacy that said one thing and did something entirely different. As the late summer days of September went by, it was clear that both France and Britain had determined – for reasons of keeping the peace in Europe – to seek compromise with Germany, and not to contest Germany's demand for the Sudetenland if the alternative was war. The sovereign territory of Czechoslovakia was to be the sacrificial lamb for 'peace' in Europe. On 15 September Hitler made his argument to Chamberlain at Berchtesgaden for the so-called self-determination of the Sudetenland, after which the British prime minister, with the agreement of France, proposed the transfer of some Czech territories to Germany. The Versailles carve-up of Europe, Britain concluded, had been a mistake and needed to be partly undone. Czechoslovakia was not consulted.

There is a horrifying inevitability about this process. Should Britain have intervened militarily to help Czechoslovakia? In 1938 Britain did not have the military capability to enable it to do this, even if it was politically possible or achievable. What Britain had failed to do from the beginning was to understand the malevolent intent behind Hitler's ideas, and oppose them politically and diplomatically, backed by the deterrence only a strong army, navy and air force could provide. The rot had started many years before in the deliberate running down of a military capability that could, for instance, have stood up to Hitler's expansionism in the Rhineland in 1936. Foresight, statesmanship, the preparation of national security on the basis of the precautionary principle (i.e. 'prepare for the worst, even if it might never happen') would have enabled Chamberlain to take much more than words to Germany in 1938 and come back with something more substantial than a piece of paper.

Appeasement is regarded by many as a moral failure. In reality it was primarily a political one. Very little could have been done to stop Hitler in 1938 and 1939 outside of the actual apparatus and the determined application of military power. Chamberlain went to Germany in 1938 and 1939 naked, and Hitler, after 1936, knew it. The political dance macabre undertaken by Chamberlain was his pretence that he was wearing the garb of a military power with the ability to exercise its prerogatives should he wish. Hitler knew full well that Chamberlain was, in fact, an emperor without any clothes. That was the folly of appeasement: not the moral posturing over far-off countries with people of whom we knew nothing, but the egregious denuding of one's own military power such that the exercise of diplomacy on a fractured international stage became nothing more than empty posturing, because it was not underwritten by *actual* power. Whatever else it was, appeasement was the delusion of believing one could exercise political and diplomatic power without having access to the deployable divisions of well-armed, well-trained troops that could have backed up words with action. Years before, President Theodore Roosevelt had said: 'speak softly and carry a big stick; you will go far'. In 1938, Neville Chamberlain did the opposite – he spoke loudly and carried only a small stick; in fact, all he carried was an umbrella.

Flying into Bad Godesberg on 22 September 1938 in his Lockheed Electra, Chamberlain was shocked to discover that Hitler now rejected the arrangement agreed at Berchtesgaden. Instead of launching plans to de-escalate tension, Hitler now ratcheted up the pressure on London and Paris by issuing an outrageously one-sided ultimatum (disguised as a 'Memorandum') demanding that the Sudetenland be ceded to Germany no later than 28 September. If the Czechs failed to see sense, Germany would intervene to 'protect' the Sudetens by force. So much for the agreement the previous week. Hitler wasn't now talking about 'adjustment' but about the wholesale absorption of part of the sovereign territory of Czechoslovakia into the Reich. The terms Hitler insisted upon, of course, all favoured Germany. Chamberlain refused. Prague, naturally, also rejected this threat, while France mobilised 600,000 soldiers, and Britain placed the Royal Navy on alert. Perversely and prematurely the German press congratulated Hitler and Chamberlain for their diplomacy, all in the pursuit of avoiding war in Europe.

It was at Munich, on 29 September 1938, that Hitler got what he wanted. Chamberlain and Daladier met with Hitler and Mussolini and agreed that Germany be allowed to take the Sudeten Province from Czechoslovakia. Chamberlain, as Ronald Blythe has observed in *The Age of Illusion*, seemed 'privately determined to confront the sordid dictators with the whole armour of a 19th-century Christian liberal English gentleman. The pity, of course, was, it was not his own other cheek he turned for the blow.'[4] He gave up the strong line he had taken at Bad Godesberg against the Memorandum. The result, as Halton observed, was a pantomime:

> We were begging Hitler to take what he wanted – if only he wouldn't shoot – and Chamberlain was being hailed as nothing less than a saintly knight-errant for doing the begging. Britain and France were imploring Hitler to take those things which would make him master of Europe and make it necessary for Britain and France to prepare for a gigantic conflict in which victory was less certain than in 1938.[5]

On 30 September, the Czechs learned that despite their own wishes France and Britain had handed the Sudetenland to Germany. There would be no war, as the Czechoslovaks had no choice in the matter: London and Paris had told Prague that if it wished to fight, it would have to do so on its own. The president, Edvard Beneš, resigned and the Wehrmacht marched. It was 'peace in our time' according to Chamberlain, who was greeted by ecstatic crowds when his Electra 14 touched down in Heston on 30 September. Churches had been packed across Britain, people praying for peace. Many in Germany also believed that Chamberlain was right and thanked God for the peacemakers. But as Shirer and others prophesied loudly and often, the sad truth was that few in the West realised that Munich, far from removing the potential for war, confirmed its imminent prospect. Czechoslovakia was not as far away as Chamberlain supposed, nor was the quarrel of no import. Either London and Paris gave in to Hitler every time the German dictator wanted to secure peace on his one-sided terms, or they would have to fight. Giving in to the tyrant on one occasion would not improve the bully's behaviour. It merely forced the democracies to choose whether to continue to give in or to make a stand. It was easy to give in when the interests they were being asked to defend were far off, and separated by vast distances of geography, culture and history. But when those interests came closer to home, the choice suddenly became much more difficult.

On 15 March 1939 Hitler completed the occupation of Czechoslovakia that he had begun with the post-Munich occupation of the Sudeten provinces. At a stroke the Czechoslovak state established in 1919 was dismembered. This action propelled Europe – and the world for that matter – into the Second World War. Those who had attempted thus far to appease Hitler now realised that they had been wrong. He could not be trusted, as his foreign policy did not follow any rational, observable course (it did, of course, if they had read *Mein Kampf* or the 1920 Programme of the National Socialist German Workers' Party or had bothered to take Hitler at his word). Thus, Britain and France were forced to offer Poland guarantees for its own security, which meant that if Germany continued its process of military aggrandisement, Britain and France were firmly and formally committed to war. It meant that Hitler now had a choice: to give up any further territorial claims or risk war with the West.

The problem for Britain was that, despite some rearmament since 1935, the country was dangerously ill-prepared for war. Vast swathes of opinion makers had fallen for the rubric that suggested that rearmament would invariably lead to war, while simultaneously divorcing themselves from any comprehension of the motives and inclinations of the users of force. Put another way, the argument was that the possession of strong and capable armed forces would invite retaliation, and therefore be the progenitor of war, not its eradicator. What the advocates of this belief failed to appreciate was that in the hands of a free and democratic nation, strong armed forces serve as a deterrent to aggression. Strong military forces in a peaceful country on the whole mean more, not less, peace, while strong armed forces in a country in which peace is not valued will invariably have dangerous consequences. A country inevitably makes itself weaker if it cannot respond to military aggression by another. To counter threats, military power needs to be sufficient to deter a potential aggressor. It is important that a rational decision-maker is never given a positive answer to the question: 'Do I think I can gain an advantage by my neighbours' military weakness or unpreparedness?' In other words, equality in the balance of power is required. This has ever been the reality in the way in which states have interacted with each other, maintaining an accepted equilibrium of power.

All political perspectives got it wrong. Conservative policy while in government disastrously miscalculated Nazi ambitions. Some in the same party saw in Nazism a European bulwark against communism. Meanwhile, the Labour Party, campaigning in the British 1935 general election, used the slogan 'Armaments Mean War – Vote Labour' and in 1938 used the similar slogan 'Stop War. Vote Labour.' It was well meaning but completely wrong, a denial of the experience of history, the reality of a century of European politics and a profoundly erroneous view of the reality of human nature. The prominent American reporter, columnist and radio personality Dorothy Thompson saw this clearly in 1934, when she observed that one of the most dangerous illusions in the Anglo-Saxon mind was 'that all peoples love liberty, and that political liberty and some form of representative government are indivisible'.[6] Many centuries of political culture in both Britain and America, where individual liberty had developed to be the fountainhead of social existence, had blinded both peoples to the reality that not everyone thought like them. In 1939 the imbalance of power in Europe was such that the aggressor nations – Germany, Italy and the USSR – were emboldened by the pacific nature of the European democracies, which smaller nations had trusted for their defence. The politics of wishful thinking would never survive in a collision with a Nazi worldview built on completely different principles.

Hitler's gobbling up of the remainder of Czechoslovakia in March 1939 proved appeasement to have been a false god. Notwithstanding the honesty of Chamberlain's motives, appeasing Hitler simply demonstrated to the Nazi hierarchy the pusillanimity of Britain's political leadership and confirmed Berlin's determination to continue on the path of national expansionism that lay at the heart of the Nazi plan – *Grossdeutschland* and *Lebensraum*. 'Our enemies have leaders who are below the average. No personalities. No masters, no men of action ... Our enemies are small fry. I saw them in Munich,' Hitler crowed to his commanders-in-chief at the outset of war on 22 August 1939.[7] Indeed, appeasement in its 1930s context – allowing Nazi Germany to achieve its political goals at the expense of the sovereignty of others, in order to avert the threat of war against themselves, or perhaps (as some have mistakenly argued*) to give Britain time to rearm – ironically enabled totalitarianism to flourish, and led to calamitous war. The problem was that those in London and Paris who pursued a policy of compromise hoped (wrongly, as it turned out) that appeasement would have no negative consequences for their own countries, even though it spelled doom for the victims.

---

*This argument does not hold water. As Chamberlain's closest adviser, Sir Horace Wilson, observed in 1962: 'Our policy was never designed just to postpone war, or enable us to enter war more united. The aim of appeasement was to avoid war altogether, for all time.' An excellent account of Wilson's role in this policy is Adrian Phillips, *Fighting Churchill, Appeasing Hitler: How a British Civil Servant Helped Cause the Second World War* (London: Biteback, 2019), p. 338.

The disaster at the heart of appeasement was the acceptance that while one was attempting to contain the dictators, bad things might happen to others, though (hopefully) not to oneself. This had been the case in respect to Japan in China in 1931, in which Manchuria was allowed to fall into Japan's expanding orbit. Britain was also to appease Japan again in 1940, over the issue of US supplies reaching the Chinese via the Burma Road from Rangoon. After all, as Chamberlain admitted to the British public about Czechoslovakia in 1938, this was about a quarrel in a far-off country between peoples of which we knew little and, by extension, cared less. It was a policy of acceptable harm. 'I don't believe myself that we could purchase peace and a lasting settlement by handing over Tanganyika to the Germans,' he observed, in explaining his approach, 'but if I did, I would not hesitate for a moment.'[8] Winston Churchill captured the problem, as he often did, with a pithy saying in January 1940, warning the European neutral powers: 'Each one hopes that if he feeds the crocodile enough, the crocodile will eat him last.'[9]

Chamberlain's approach was predicated on the firm belief that, by logical and reasonable discussion and argument, Nazi minds could be changed. The great failure of the 1930s was that Europe (and the United States) refused to acknowledge the rise of fascism for what it was – not merely a threat to the liberty of citizens and the rule of law *within* a state, but the end of the rule of law *between* states and a reversion to the unchecked brutalism of power-based politics – and to do anything practical about preventing its unimpeded growth. This was a failure to defend Britain's national interest, because it made Britain less, rather than more, secure. Allowing Germany and Italy to do what they wanted on the international stage directly impacted British interests, security and otherwise, and provided a specious solution to an ongoing political problem. The rise of totalitarian politics determined to use brute force to secure the dictators' policy aims rather than operating within the 'rules-based order' (to use a more modern phrase) was nowhere adequately confronted and denied. It was allowed to march unchecked for fear that action against it might itself lead to war. There were few voices who advocated pushing back against Germany and Italy and adequately preparing for war. It was the fear of war that perversely allowed war and violence to flourish. Germany's political trajectory was repeatedly excused, and the consequences of Nazi ideology denied or ignored.

The robust *Pax Britannica* of an earlier era, in which Britain stood up firmly for its national interests globally – which in the previous century included a sacrificial war against slavery led by the Royal Navy – was replaced in Europe in the 1930s with a much-derided (in Europe) *Pax Umbrellica*, in which Chamberlain's badly furled umbrella was seen to signify weakness, equivocation and uncertainty. Mussolini scoffed that 'people who carry an umbrella can never found an empire'.[*] Thus were the impulses of appeasement bred. It needs to be emphasised, however, that Chamberlain's approach – in which he genuinely believed that war could be averted by personal diplomacy – was wildly endorsed across most sectors of opinion in Britain, and across the Dominions, concerned never again to become embroiled in another bloody and pointless European imbroglio. *The Times, Daily Mail, Daily Express* and *Observer* were all noisy proponents of appeasement, as was, overwhelmingly, the BBC, the Church of England, the House of Lords and the City of London.[†] It made sense to think thus, if the alternative was war. The problem was that the political and psychological basis of appeasement, as events were to show, was fatally flawed. At the time of Munich in September 1938, for instance, the Wehrmacht had only three tank divisions and enough ammunition for a six-week campaign. By not starving it, Britain and France allowed the crocodile to become fatter and more aggressive. War was made

---

[*]D.R. Thorpe, *Eden: The Life and Times of Anthony Eden, First Earl of Avon 1897–1977* (London: Chatto & Windus, 2003), p. 193. If Mussolini had seen the Naga tribesmen of Mokokchung in Assam carrying black umbrellas and singing 'God Save the King' on the return of the Assam Rifles from a punitive expedition against head hunters in 1936, he might have changed his mind.

[†]Though Tim Bouverie in *Appeasing Hitler: Chamberlain, Churchill and the Road to War* (London: The Bodley Head, 2019), p. 417 notes that at the time of Munich 43 per cent of Britons were opposed to the appeasement of Hitler, with only 22 per cent in favour.

more likely, rather than less, by appeasement. Many opportunities to stop the growth of Nazi ambition were ignored, each additional month of well-meaning but wrong-headed diplomacy giving Germany the time it needed to strengthen its rearmament.

The exchange at Munich was that Europe was promised a wider peace in exchange for the enslavement of Czechoslovakia. It was a chimera, caused by the failure of Britain to defend the peace established at Versailles with the threat and, if necessary, the use of force, in conjunction with France, its Great War ally. The tragedy of 1938 was that the price of the exchange was not paid for by those making the agreement (Britain and France on the one hand, and Germany on the other) but by Czechoslovakia, the victim. The agreement, dressed up by Chamberlain as delivering 'peace in our time', was ultimately deceitful, because it was written on a cheque from a bank in which London and Paris did not even have an account.

PART FOUR

The end of the beginning

Chapter 15

The empire declares war

During the summer of 1939 the German government thundered threats against what they claimed to be the grotesque mistreatment of German interests by the criminally minded government in Warsaw. How many innocent Germans were fated to die cruel and barbaric deaths at the hands of illiterate, uncivilised Poles while Germany stood idly by? Wodan's war drums hammered their insistent beat, loudly seeking revenge for the humiliation of Versailles: the wicked treatment of the German minority by the Poles, which was caused by the illegal division of Europe in 1919, was the *casus belli*. The emasculation of Germany after the Great War had been an outrageous piece of revenge politics, so went the argument, which had resulted in millions of pure-born Germans being subjected to low-born slave masters in an oppressive Slav state. Natural justice demanded that this intolerable situation be reversed. Berlin was offering the Poles the opportunity to do so voluntarily. If they didn't accept Berlin's generous offer, the threat was clear. The manufactured hysteria describing 'Last Warnings', 'Unendurable Outrages' and 'Murderous Poles' had been splashed over the front pages of the German newspapers for much of the year. There is no doubt that the Polish mobilisation in April had exacerbated the mud-slinging, although by late August most newspaper-reading Germans were exhausted by this talk of war.

On 11 August Bill Shirer visited the Free City of Danzig (Gdansk). Despite the huge volume of military traffic on the streets of the old seaport he reported that the local inhabitants believed that Hitler's bellicosity was specifically designed to avoid war, not to create it. The people certainly wanted to be allied to Germany, but not at the expense of war or the loss of Polish trade. Shirer was less sanguine. His view was that Poland – like Czechoslovakia before it – was threatened with extinction as long as it presented itself as a threat to a resurgent Germany. Hitler simply could not have a strong Poland – or a strong Hungary, Romania or Yugoslavia, for that matter – on his eastern border. Berlin's demand to regain sovereignty of Danzig in the cause of pan-Germanism was partly a pretext for subjugating the threats against it to the east, but also a genuine expression of the long-term Nazi demand for *Lebensraum*.

On 23 August the entire world changed. The Nazi–Soviet Pact was announced to the amazement of everyone outside of Hitler's inner circle, though most of its contents remained a closely guarded secret. In Germany, the Soviet Union had always been the 'evil empire', and to find one's country allied to its erstwhile and bitter enemy was, at the very least, confusing. But it was immensely popular, especially with the traditionally left-leaning working classes, principally because the people assumed that it meant the absence of war on two fronts, ever the haunting spectre of German strategy. To the consternation of the German populace, two days later the British signed a treaty of mutual assistance with Poland.

In London at the end of the month, Matthew Halton found himself walking with Viscount Cecil through St James's Park. It was a delightfully bucolic scene. He recalled:

Ducks squabbled under the little bridge; flamingos poised themselves on one leg with disdainful grace; pelicans gazed solemnly at the white towers of Westminster Abbey. Children raced between the thick velvet lawns and bugles rang from the nearby barracks of the Household Cavalry. Visible over the Admiralty Arch, Nelson squinted from his pedestal in Trafalgar Square; and on Horse Guards Parade water bubbled from the little memorial fountain with its inscription, 'Blow out, ye bugles, over the rich dead'...

The conversation with Cecil was about the epitaph that would soon be written to the disastrous end to two decades of the failed policy of appeasement. 'Perhaps the coming war will teach the world,' he mused, 'that mankind's first line of defence is its conscience, and that principles are more potent than guns. Hitler can be defeated now only by guns. It may teach us to put force in the hands of law.'[1]

Not everyone agreed or could see the future with such perspicacity. Right up to the declaration of war, the balance of opinion among the British General Staff was that it wouldn't happen. At the War Office, Major General John Kennedy recalled on Friday 25 August 1939 that odds were taken that morning on whether or not there would be a war. 'Gort offered 5 to 4 against; [General Sir Ronald] Adam, the Deputy C.I.G.S., 6 to 4 against; Ironside was laying 5 to 1 on. The division of opinion continued right up to the actual outbreak.'[2] London's understanding of what Berlin might or could do, was shockingly limited.

On 31 August Hitler gave the order for his troops to cross the Polish frontier the following day. In accordance with Britain's promise to Poland, at 11am those people across Europe who were tuned in to the BBC heard the slow and uninspiring tones of the British prime minister announcing that Britain and Germany were now at war. Chamberlain was glum, his hard-fought policy of appeasement crashing around him. Within minutes the first air-raid siren began to wail over London, sending millions of people scurrying for cover. It was, of course, a false alarm. Such was the hype that had been whipped up about the nature of modern warfare that many believed that the clouds would be instantly dark with enemy bombers. Another alarm was sounded that night, the noise incongruous with the beauty of the night sky, stars twinkling in the firmament far above the armada of vast grey, bloated barrage balloons that sat above the city as a deterrent to low-flying enemy aircraft. 'London on a moonlit night in the blackout was a place of fantasy,' Halton remembered. 'Thousands, perhaps millions, of

Londoners saw the stars almost for the first time in their lives.' Someone remarked to Quo Tai-Chi, the Chinese ambassador in London, that the sky would soon be dark with enemy bombers. 'The sky is already dark,' Quo Tai-Chi replied, 'with the wings of chickens coming home to roost.'[3]

At the beginning of 1939 there were 107,000 regular soldiers in Britain. The immediate shortfall of troops needed for the BEF was made up from men called up from the reserves (i.e. ex-soldiers, who had a reserve liability), many of whom had been out of uniform for several years, and the remainder from the Territorial Army. After this came those conscripted from April 1939. The army that Britain began to mobilise in April 1939 following the announcement of conscription was a far cry from the army it had fielded in 1918. The victorious, purposeful, well-drilled and combat-focused army of 1918 was large – a citizen army that had been transformed into a professional organisation able to master the intricacies of the battlefield to an extent that astonished the veterans of 1916. The rapidly expanding army of 1939 had to build on the emasculated structure bequeathed it in 1938, with an impoverished concept of warfighting, and little residual corporate knowledge about how to fight on a fast-flowing manoeuvre battlefield. Consequently, the army sent to France in 1939 was little more than a vast conglomeration of ill-equipped citizens in uniform bound together more by a reluctant sense of the inevitable than by a defining methodology or philosophy of battle.

When appointed to lead the BEF, Gort was exhorted by the new CIGS, General Sir Edmund Ironside, to get it safely to France and dig in. It was 1915 all over again. There would be no foolish race forward to Mons, to engage in an encounter battle à la 1914. Instead, the trenches of 1915 seemed to offer the only hope not merely of stopping a German offensive, but of protecting this newly enlarged and untrained army from annihilation. The great strength of the rapidly expanding army nevertheless remained its regimental structure, which provided small, perfectly formed families in which newly joined conscripts could find a home. The army retained in its ranks many who still wore on their chests the medal ribbons of the Great War and whose battalions and regiments carried its battle honours. At the very least they knew that when the chips were down, no matter how poorly prepared or organised, the army could *fight*. This was the great tragedy of 1939: Britain had had the opportunity after 1919 to retain and further develop the kernel of an army able to fight and win against a peer adversary but had grievously squandered the opportunity. It had tilted at the beguiling windmill of empire, arguing to itself that the Great War – or the European War as it was often described at the time – could never be repeated. And yet, here we were again. Once more Britain and its army would need to build itself from the ground up and learn what was required to prevail in a battlefield it had ignored for two decades. It would be able to do so, however, only if the enemy gave it time to recover and rebuild.

The fast-growing army, both regular and Territorial, plundered the warehouses of boots, webbing and uniforms left over from the Great War. There were shortages of everything. The first problem, however, was that of Britain's manufacturing capacity. The vast industrial manufactory of 1917 and 1918, so great that it was able to replace everything lost in the German March 1918 offensive within a matter of weeks, had been rapidly dismantled after 1919 in the not unreasonable rush to turn swords into ploughshares. Rifles, Bren guns, 2-pounder anti-tank guns, 3.7-inch anti-aircraft guns, armoured cars and tanks, together with a million other pieces of military equipment, had to be produced rapidly, with a dramatically curtailed level of industrial expertise, and factory space, available. An audit in July 1939 noted that of a requirement for 240 anti-aircraft guns required for the BEF, there were only 72 available, and of the 240 2-pounder anti-tank guns required, there were only 144. There were none of the new 25-pounders for the artillery, the only field artillery available being 18-pounder veterans of the Great War. Only 60 of the 1,646 Infantry or 'I' tanks required had been built by August 1939, none of them the Mark II (Matilda) version. Tank battalions and armoured reconnaissance regiments had to make do with the three-man Carden Lloyd Mark V light tanks built by Vickers. In any case, as has been seen, there was no coherent articulation of armoured warfare in the British Army in 1939 before the start of the German offensive. The effective deployment of tanks in battle remained a fiercely fought over and contentious issue that had not resolved itself in a formal doctrine of how armour was to fight.* What compounded the problem was that by not having a clear idea of how armour could be used in a manouevrist way, when the panzer divisions struck in May 1940, the response was shock and surprise, and near-disastrous tactical responses.

The second problem was formation training. The newly recruited masses were quickly put through their basic training, enabling them to drill, dress and salute smartly, organise their personal kit and bull their boots. Some minor tactics and field craft were learned, but it was rudimentary and was not extended to all. The extent of personal training was, for most infantrymen, skill-at-arms on a rifle range, though an abject shortage of both .303 ammunition and rifle ranges curtailed even that. If troops learned anything, it was how to dig and revet a trench, and lay out wire entanglements.[4] Officers trained with sandpits and TEWTs (Tactical Exercises Without Troops), which lacked the necessary realism to enable newly commissioned platoon commanders to grasp what they would have to do to manage their platoons when under the stress of incoming fire. For most of the time military training was substituted for guard duties, or undertaking unskilled manual labour, a problem exacerbated on arrival in France, when the accumulated shovel-power of

---

*Indeed, it was not to have one for several years yet. See J.P. Harris and F.H. Toase (eds), *Armoured Warfare* (London: Batsford, 1990), p. 29 and J.P. Harris, *Men, Ideas and Tanks: British Military Thought and Armoured Forces, 1903–1939* (Manchester: Manchester University Press, 1995).

the BEF was concentrated on digging an enormous array of trenches along the nearly 30 miles of British front, nicknamed the Gort Line. The idea of battle schools, where soldiers learned to prepare for the trials of combat through rigorous inoculation, including the use of live rounds, was still in the future, although some divisional commanders, such as Major General Harold Alexander, were keen proponents of standardising a common battle drill across the army. As has been seen, sustained and rigorous formation (brigade and above) training was absent for all of the 1930s, the formation exercises that did take place being short and hopelessly unrealistic.

In any case, at the outset of the war the British did not consider that the Germans held any particular military advantage. Major General Sir Henry Pownall noted in his diary on 1 April 1940 that, because of the work the Belgians had 'done on their defence ... Hitler's prospects of a rapid break-through in Belgium are surely much smaller than they were. We ought to give him a proper knock if he tries it on.'[5] Trenches would stop the Germans, just as they had in 1915. The army had forgotten 1918 entirely and, because it had not expanded doctrinally on the lessons of the Hundred Days, believed implicitly in Clausewitz's dictum that the defensive was the most important operation of war. In any case, as Major General John Kennedy observed: 'A dictum of Foch's was still widely believed; that the tactics of each war began where those of the last left off; and it was deduced from this that we might expect an initial period of trench warfare in which to build up bigger British forces.'[6] Even British soldiers had forgotten that 1918 had not ended with trench stalemate, but with a form of (Allied) *Blitzkrieg* against the Hindenburg Line. Accordingly, in 1939 Britain was starting from a very low baseline of knowledge and expertise in the art of modern war. Kennedy noted:

The controversy between Ministers and the General Staff dragged on wearily, confused by many red herrings and false prophets. Thus, the Maginot Line was pronounced impregnable; the attack in modern warfare could not possibly prevail over the defence; air forces alone would suffice to stop the advance. Those years were the heyday for cheapjack purveyors of bogus military nostrums.[7]

On 3 September 1939 the British Army found itself stretched across the globe. In addition to the rapid growth of the regular and Territorial armies at home and the imminent despatch of the BEF to France, the army was in Egypt (one armoured division in the process of forming, four infantry brigades including one from Palestine, together with additional Indian artillery and engineer units); Palestine (Headquarters 7th and 8th Infantry Divisions and three infantry brigades); East Africa – Sudan, Kenya, Uganda and British Somaliland (two British battalions and five infantry brigades (i.e. 15 battalions) recruited in East and West Africa); and India.[*] The 190,000-strong Indian Army had been considered, since 1933, as part of the empire's strategic manpower reserve, in order to relieve the pressure on the fast-dwindling and cash-strapped British Army and as a cheap way of populating British overseas garrisons. But the Indian Army was a poor relation to the British Army in terms of equipment and of training for a modern war. It had no armour. It was only in 1935 that artillery was reintroduced to the Indian Army as an Indian combat arm. With the war clouds gathering over Europe in 1938 a modernisation programme sought to bring the Indian Army rapidly into the modern age, designed to fight alongside the British Army and against a so-called 'first-class' opponent. Doing nothing would mean that the army would 'fall behind the forces of such minor States as Egypt, Iraq and Afghanistan'.[8]

If the Indian Army was to play a role in support of the British Army, potentially in Europe as had happened in the Great War,

---

[*]There were 56,000 troops in India, 14,000 men scattered in garrisons across the far-off empire and 21,000 in the Mediterranean and Middle East.

change was imperative. A deployable force was created, later to be designated 4th Indian Division (commanded by Lieutenant General Francis Tuker between 1941 and 1944), but the journey to change the Indian Army into a modern force, trained to fight confidently abroad, alongside and equal to formations of the British Army, with all the equipment and materiel necessary to enable success, was clearly going to take considerable time and effort. As the committee led by General Chatfield to evaluate what was required concluded in 1938, the Indian Army's equipment was poor, relying on the cast-offs of the British Army. For example, Indian battalions at the time possessed only 12 Vickers-Berthier light machine guns, one per platoon of 40 men. By contrast, the Japanese had a light machine gun in each section of 12 men. The shocking state of the (dis)armament of the deployable Indian Army in 1941 and 1942 is only now vaguely recalled but is of paramount importance in assessing the fighting effectiveness of units when facing the Japanese juggernaut in 1942.

Nevertheless, the modernisation programme, led by the commander-in-chief, General Sir Claude Auchinleck, was to have an unexpected bonus, for it established the mechanisms for the rapid growth of the Indian Army when the requirement came at the outset of war in the Far East and Burma. Within a year of the declaration of war, the army had grown four-fold, volunteers flocking to the call to arms from a far wider section of the population of the subcontinent than had been the traditional recruiting ground of the Indian Army, but at a rate that far outstripped the ability of India to produce the weapons and warlike stores necessary to equip them, or the instructors necessary to train them. Newly formed infantry battalions could expect to receive just about enough rifles to issue to recruits, but only a fraction of the automatic weapons required and no mortars, grenades, Bren guns or Boys anti-tank rifles. General Sir Archibald Wavell, Commander-in-Chief India, re-emphasised this running sore with a note to London on 11 September 1941, two years after the start of the war in the West, that India possessed only 20 2-pounder anti-tank guns. The fact was that Britain didn't have many more, and couldn't equip its own army at the time, let alone India's. By the time of the Japanese entry into the war in December 1941, the army had two weak divisions in Burma and one Australian and two Indian divisions in Malaya, together with a Canadian brigade, a British brigade and an Indian battalion in Hong Kong.

The confusion of getting the BEF to France and into some semblance of order along the Gort Line in late 1939 was almost overwhelming. The BEF had deployed, but how would it fight? How would the Germans attack, where and when? Nothing was certain. The Air Staff continued to push for a strategic air offensive against the Ruhr. Major General John Kennedy, at the heart of planning in the War Office, despaired. 'We studied plans and hypotheses; we churned out notes, briefs and memoranda on every conceivable subject. Never had so little fighting produced such a voluminous range of documents.'[9] Clarity remained absent, however.

First blood was spilled for the British Army in a disastrous affair in Norway in April 1940. There was little about the operation that redounded to the credit of the armed forces generally or the army specifically. The story is simply told. With Britain's eyes on the prospect of an attack against France and the Low Countries once Germany had finished with Poland, the German invasion of Denmark and Norway in April 1940 came as an unpleasant surprise. Britain was already exercised about German access to Swedish iron ore and had considered sowing mines across the Baltic. But while London thought through its options, Germany moved, seizing Copenhagen, Oslo and the major Norwegian ports. It was entirely unexpected in London. As he struggled into his clothes on the morning of 9 April after receiving news of the German attack, General Ismay considered, he recalled, 'for the first time in my life, the devastating and demoralising effect of surprise. I had always thought that Hitler's next move would be either an invasion of the Low Countries and France, or alternatively an air attack on the British Isles.'[10]

Norway now beckoned as the first battlefield in this new war with Germany. The first problem was that there was no one in the War Office with any idea as to what to do. There had been a plan the previous year to send an expeditionary force there to transit through to Finland. The Finnish surrender in March 1940 had put paid to that idea. An Anglo-French plan to mine Norwegian waters to prevent the transport of iron ore to Germany, and to land forces in the country (against the wishes of its government, Norway being neutral) to enforce this, in order to prevent Germany from using Norway's fiords and harbours for its navy, especially its U-boats, was to be put in place in early April. Hitler forestalled this with impeccable timing, leaving the Allies scrabbling for a response.

It was decided that the Royal Navy would attempt to remove the Germans from Bergen and Trondheim, while the army – and the French – sent troops to Narvik, far to the north, to prevent the Germans conquering the whole country and securing its ports for the prosecution of a sea war against Britain. But as with all things half-planned, with ill-prepared and negligibly trained troops, the operation became unstuck from the start. The Germans, with starter's advantage and very significant air superiority, put the bumbling Allies to shame.

A force of four British infantry battalions in 24th Guards Brigade together with French *Chasseurs Alpins* would be sent by sea to occupy the port of Narvik. This force would prevent German occupation of the port and allow Britain to retain a pivot for future action in the very far north of the country. The earliest that General Edmund Ironside, the CIGS, could promise for the landing of these hastily gathered troops was 23 April. The problem was that the Germans got there first, troops being landed by a flotilla of Kriegsmarine destroyers in mid-April. At the same time, news arrived that the Norwegian government, which had evacuated Oslo at the start of the invasion, had decided to resist the invader and wished to continue the fight from Trondheim in central Norway.

The idea was now developed to land troops totalling 23,000 men in both northern and central Norway. Nothing of this kind had ever been planned before, and few British politicians or soldiers understood anything about the country. It was clear that the distance between the two locations meant that neither operation would be able to support the other. Accordingly, two operations were hastily cobbled together with different commanders. Then, in mid-April in two separate actions the Royal Navy proceeded to destroy the ten Kriegsmarine destroyers in Narvik, leaving the German troops in the town – of unknown quantity – cut off. This persuaded London that an attack into the north, which few had any inkling was still lying under feet of winter snow, would only require the Guards Brigade. The French would be diverted to support the force heading for Trondheim. The commander of this operation was to be Major General Carton de Wiart, the recently appointed general officer commanding 61st Division. He wasn't to take his division, however, but would take command of 146th Brigade, a Territorial unit mobilised at the outbreak of war.

But knee-jerk – and even panicked – political decision-making about what to do about the sudden and unexpected problem in Norway was combined with ad hoc planning and deployment of forces. There was no single campaign plan. Nor was there any intelligence about German plans in Norway, nor of the state of their forces in Narvik or those approaching Trondheim. Events in the country were fluid and intelligence was poor. What was the strategic purpose of both operations to be? Was it to hold Narvik and defend Trondheim from German attack indefinitely? If so, how could these forces be sustained? Who – and what – would provide air cover? What happened if the Germans occupied Trondheim first: would the role of the army be to recapture it? For what, and why? A lack of careful consideration in respect of grand strategy ('what is the strategic purpose for Britain's military involvement in Norway, and how will a military operation enhance Britain's security?') was accompanied not merely by an almost complete absence of command and control at the operational level (i.e. the campaign level), but by a series of cock-ups that doomed the operation from the start.

The British Army and Royal Navy had not operated together in the transhipment of troops in an active operation like this since Gallipoli in 1915 (itself a disaster at the operational level of war). The benighted 146th Brigade, rushed to the dockside to load onto two Royal Navy cruisers, succeeded only in loading all their kit and equipment before the destroyers were suddenly ordered to rush out to the North Sea, leaving the men behind. Although the brigade was able to land at Namsos, north of Trondheim, on 17 April, they did so without their equipment, and without their commander, Brigadier Charles Phillips, whose ship was diverted to Narvik instead. To make matters worse the Luftwaffe arrived on the scene during the Allied landings, destroying the town of Namsos and most of the disembarked stores for the French *Chasseurs Alpin* in the process. Carton de Wiart's Sunderland flying boat was attacked by a German fighter and forced to ditch, de Wiart being rescued by a Royal Navy destroyer. To his chagrin he observed when ashore that there were no anti-aircraft weapons to fire back with, nor any Allied fighters to secure control of the skies. Nor did the disembarked French arrive with any transport to move them south to Trondheim. They would have to move themselves and their equipment by foot.

The Germans promptly undertook successful operations to destroy the landings. They had air cover, artillery, transport and a plan. Carton de Wiart had none of these things. Namsos quickly became an embarrassment of inflated ambition combined with deficient planning and inadequate resources. Carton de Wiart told London so, but was ordered to remain on the defensive. Only on 27 April was it accepted in London that the Namsos force would need to be evacuated. Ironically, the evacuation between 29 April and 1 May was effectively and competently managed by the troops on the ground co-operating with the Royal Navy, though virtually all their equipment was left behind. Brigadier Phillips:

> summed up the reasons for failure. He noted that no advance plan had been made, that he had no maps of the area in which he was to operate, no snowshoes, no skis, no anti-aircraft guns,

little artillery, gunners who had never fired a shot in anger and in any case had no ammunition to fire. He had troops with no transport, many of whom had never fired a rifle or a mortar. They had received no brigade training, had no air support and no communications, having been forced to use local Norwegian telephone exchanges. 'Finally I am forced to the sad conclusion,' he ended, 'that not only the lessons we learnt so dearly at Gallipoli, Salonika and Mesopotamia have been forgotten but we now make even more mistakes of a most elementary nature.'[11]

The disaster at Namsos was replicated in an equally disastrous landing by a battalion and a half (1,200 men) of 148th Brigade to the south of Trondheim at the small port of Åndalsnes. To those in the armchairs in London looking at the map of Norway, it appeared that Trondheim could be secured in a pincer, one from Namsos in the north and the other from Åndalsnes in the south. The commander of 148th Brigade – Major General Harold Morgan – likewise landed without any of his equipment, his motor transport, four Bofors anti-aircraft guns, 25,000 rations and a huge amount of ammunition being sunk en route. Under relentless Luftwaffe attack 1,200 men were quickly reduced to 400. Just when it became clear that an evacuation was required, London reinforced failure by landing a further brigade – 15th Brigade, commanded by Lieutenant General Bernard Paget – though likewise, it disembarked naked, without any artillery, transport, anti-tank or anti-aircraft guns or effective air cover. Paget had scoured the Charing Cross bookshops in London in an attempt to find a map of Trondheim, as he couldn't get one from the War Office. It was almost a comedy of errors, though men were losing their lives as a result of an extraordinary series of ill-thought out plans and blunders. The remnants of both battered brigades were evacuated between 28 April and 2 May. Likewise in the far north prevarication and argument between the army and navy commanders led to indecision and delay, Narvik finally falling to a mixed force of Norwegians and French. Nothing had been achieved in Norway except to highlight the utter inadequacy of British operational planning and preparation for war.

Norway 1940 was an expensive lesson in the need for armed forces to operate in expeditionary campaigns only with clear plans, air cover, well-trained and well-equipped troops with the weapons to do the job asked of it. None of these were evident in 1940. It was hardly an auspicious start for what was to come next.[12] The disastrous Norway campaign remains an object lesson to this day of how not to conceive, conduct and command a joint service operation.

Chapter 16

The chickens come home to roost

In France, Britain didn't even have a commander nominated for its expeditionary force until the day war was declared. Lord Gort, who was Chief of the Imperial General Staff and a much-decorated Great War junior officer, contrived to get himself appointed to this critical senior field command. Field Marshal 'Tiny' Ironside had expected the role but was instead appointed to replace Gort as CIGS, an appointment to which he regarded himself as temperamentally unsuited. Having secured command of the BEF for himself, Gort was given the following instructions:

1. His Majesty's Government have decided to send a Field Force to France and to entrust its command to you. The role of the force under your command is to cooperate with our Allies in the defeat of the common enemy.

2. You will be under the command of the French Commander-in-Chief 'North-East Theatre of Operations'. In the pursuit of the common object, the defeat of the enemy, you will carry out loyally any instructions issued by him. At the same time, if any order given by him appears to you to imperil the British Field Force, it is agreed between the British and French Governments that you should be at liberty to appeal to the British Government before executing that order. Whilst it is hoped that the need for such an appeal will seldom, if ever, arise, you will not hesitate to avail yourself of your right to make it, if you think fit ...

4. It is the desire of His Majesty's Government to keep the British Forces under your command, as far as possible, together. If at any time the French Commander-in-Chief 'North-East Theatre of Operations' finds it essential for any reason to transfer any portion of the British troops to an area other than that in which your main force is operating, it should be distinctly understood that this is only a temporary arrangement, and that as soon as practicable the troops thus detached should be re-united to the main body of the British Forces ...

8. Whilst the Royal Air Force Component of the Field Force is included under your command, the Advanced Air Striking Force, which will also operate from French territory, is an independent force, under the direct control of the Air Officer Commanding-in-Chief, Bomber Command, in the United Kingdom. The War Office has nevertheless undertaken the maintenance of this force from the common bases up to railhead and for this you, as Commander-in-Chief of the Field Force, will be responsible.[1]

Britain's pre-war policy of limited liability with respect to the prospect of war in Europe meant that no grand strategic conversations had been undertaken with France about its defensive plans in northern France and along the Belgian frontier. There had therefore been no opportunity to shape the defence of northern France in a way that was not dependent on linear defences. The conversations that took place in 1939 were about *numbers* of troops. The criticism of France should not be exaggerated, however. Britain did not have the doctrinal understanding or vision to be able to present to the French High Command an alternative approach to defensive dispositions in north-western France and Belgium based on *how* the Germans were likely to attack. A fully formed doctrinal position within the British Army about how to counter a mobile offensive based on the lessons of 1918 simply did not exist in 1939, meaning that a conversation with France about an intelligent alternative to the linear defence of territory could not have occurred. The French defensive strategy based on the concrete and guns of the Maginot Line prevailed.

The British had agreed, in April 1939, to increase their contribution to the defence of France from two to four infantry divisions. In retrospect the numbers seem laughably small, and late. These troops, organised in two corps (one commanded by Lieutenant General Alan Brooke and the other by Lieutenant General Sir John Dill), crossed the Channel in September 1939. By the end of September 160,000 troops were in France, and by the time the Germans struck in May 1940, the BEF had near-quadrupled to over 380,000. The greatest deficiency lay in the preparation of senior officers for command, and the training of formations. Gort had only ever before commanded an infantry battalion. By the time of the German invasion in May 1940, he commanded 13 divisions. To give some sense of the scale of the British contribution to the defence of France, the French had 110 divisions and the still-neutral Belgians, 16. Confronting them the German Armies in the West had 134 divisions, of which ten were panzer divisions.

Likewise, a very considerable deficiency in the British Army lay in the lack of adequate formation training for war. There had been no mechanism in the inter-war years for senior officers to be trained as brigade-, division-, corps- or army-level commanders. Moreover, the failure of battalions, brigades and divisions to train in realistic, demanding exercises over considerable tracts of land, and to train together with all arms working together against a common defensive or offensive plan, and with aircraft was principally due to budgetary constraints. However, it was also down to a lack of imagination by most senior officers to envisage what might be necessary, and meant that no divisions had ever seriously trained as an entity, nor with other divisions. There were exceptions; Lieutenant General Alan Brooke (II Corps) and one of his divisional commanders, Major General Bernard Montgomery (3rd Infantry Division), tried hard in France to train their men for the forthcoming battle. But it was too little too late.

Moreover, it went further than a lack of training. The greater problem was the lack of a doctrinal template, especially at divisional level, against which to train. How should a complex defensive battle be fought against multiple attacking armoured thrusts? How should an offensive be mounted, utilising the full combat power of the army, to beat the enemy in both time (i.e. thinking and acting faster than the enemy) and space (i.e. using terrain and air space to their maximum advantage)? Apart from establishing dug-in and wired defensive lines, the British Army had no idea about the former and, because it had not embraced and developed the lessons of the Hundred Days campaign in 1918, had even less notion about the latter. In the first place, it couldn't conceive of what was about to hit it and, because this was beyond its imagination, it had no counter except for the bravery and steadfastness of its men. It was not the men of the BEF who were defeated in 1940, but the paucity of its military *thinking* that meant that its *doing* proved to be wholly inadequate for the task it faced. It was no way to prepare for or fight a modern war.

During the period of grace given the Allies leading up to 10 May 1940 the BEF expanded to 13 divisions* as the Territorial Army was mobilised. This additional time proved, in the end, to matter for nothing as in its training, equipment and deployment it remained unequal to the task, as the Wehrmacht was shortly to demonstrate. It spent the winter and spring shaking itself out following the rapid increase of the army, receiving vehicles, building up depots to support the front line and developing the fixed defences nicknamed the Gort Line – an anti-tank ditch, trenches, pillboxes and barbed wire – along the 28 miles of its allocated part of the line, between Maulde and Halluin, on the Franco-Belgian border. The idea that the British could have spent this time training instead for a mobile war was a nonsense: the BEF wouldn't have known where to start. During the eight months before the German attack Headquarters BEF undertook not a single exercise in preparation for war. Digging defensive positions and organising the arriving troops took priority over training. In any case, it was the length and strength of the

---

*This included the 51st (Highland) Division which was attached to the French Army on the Saar Front and was not involved in the operations of the BEF.

defensive line which obsessed British commanders, keen to ensure that there were no gaps through which a German attack could penetrate. The concept of creating counter-penetration or counter-infiltration positions, in great depth, was entirely alien to the BEF in 1940, simply because it had not studied, understood and grasped the power of a combined-arms, armour-led, offensive strike of the sort that was about to hit it amidships, or that it had itself deployed in the Hundred Days offensive of 1918. Twenty years of amnesia was to prove to have huge consequences.

Three divisions of Gort's 13 were not meant to fight: they had been sent to France to build airfields. The 1st Armoured Division with its 250 tanks, half of which were the woefully inadequate Vickers Mark VI light tanks armed with twin machine guns, arrived after active operations began and played no significant part in the campaign. Unlike the German Army, the only tanks in a British division were the light tanks of the divisional cavalry regiment. The 1st Army Tank Brigade, responsible not for operational manoeuvre but for infantry support, contained one regiment – the 4th Royal Tank Regiment – with 50 Infantry Tank Mark Is, armed only with a .5-inch machine gun. In early May, just before the German attack, the second regiment – 7th Royal Tank Regiment – joined the BEF, with a total of seven Vickers light tanks, 27 Mark Is, and 23 of the new Matilda Mark IIs, which boasted an excellent 2-pounder gun. The problem remained as to how the tanks were to be used. Unlike the Germans, who used theirs in 1940 for shock action, the British and French spread their armour across their front, instead of concentrating it for counter-penetration, counter-attack or combined-arms assault tasks.

The story of Operation *Sichelschnitt* ('Sickle Cut') is so remarkable that it has found its way into common mythology as a brand new type of warfare – *Blitzkrieg*. The Germans launched their attack on the Netherlands, Belgium and France on 10 May with three simultaneous offensives. In the first, General von Bock's Army

Group B, with three panzer, one motorised and 24 infantry divisions, attacked Belgium. This was what the Allies largely expected, but had nevertheless been ill-prepared for, given that Belgium's neutral status had allowed for little reconnaissance or preparation. It was in conception a re-run of the 1914 Schlieffen Plan, with the addition of the capture of the Netherlands. In the second, General Gerd von Rundstedt launched his Army Group A into north-west France through the Ardennes, with seven panzer divisions (70 per cent of the German total), three motorised and 34 infantry divisions, to cut off the Allies north of the Somme by capturing Amiens and Abbeville. In the third, the 17 divisions of General von Leeb's Army Group C were deployed against the 50 French divisions guarding the 200-mile-long Maginot Line. Both Army Group A and B were designed to work in parallel with each other; the northern offensive wasn't a feint for the fast-moving southern one as some have assumed. Both attacks were designed as simultaneous offensive manoeuvres against fixed defences, in the north Belgian and in the south, French.

In the months immediately preceding the German offensive, the Allies (i.e. Britain and France) had become increasingly worried about the vulnerability of north-western France to a German attack through Belgium. The obvious German attack route, they considered, would follow the approach adopted in 1914. They agreed, after much deliberation, that the 23 divisions of General Gaston-Henri Billotte's 1st Army Group (which included the BEF) would deploy into Belgium in the event of hostilities in order to reinforce the Belgian Army and offer a counter to the expected German main offensive.* The French Seventh Army would then advance further north to the support of the Dutch. This meant giving up the elaborate, fixed defences the BEF had prepared in France in exchange for new, unreconnoitred and unprepared positions along the River Dyle just to the east of Brussels – 'Plan D'.

---

*The 1st Army Group comprised 21 divisions of the BEF and the First, Second, Seventh and Ninth French Armies.

Between 10 May and 4 June 1940 the Wehrmacht stormed into France and the Low Countries in a decisive set of operations that were to eject the BEF from Dunkirk and cause the collapse of the French Army, the loss of Paris and the surrender of France. In the south Army Group A was able to use superior force to break through the weak French defences along the Meuse at Sedan. Striking hard against an enemy which had confidently expected the primary German attack to flow through Belgium to the north, a classic outmanoeuvre operation was developed, forcing great splits in the French defences and leaving the BEF no option than to withdraw to the Channel coast. Neither by training, nor preparation, nor expectation were the British able to anticipate or respond to this expression of operational art in 1940, something they had successfully mastered in 1918.

This left much of the area into which von Rundstedt's Army Group A rushed denuded of planned defences and defended by mobilised French reservists. The BEF constituted only 10 per cent of the French Army and, in the light of the instructions given by the Secretary of State for War to Gort, he had no choice but to conform to the overall French plan.

In retrospect the relocation of the BEF to Belgium was a strategic error that facilitated von Rundstedt's Army Group A's success. Belgium, having carefully protected its neutrality since 1937, suddenly, with the advent of hostilities, became an ally of Britain and France again. Accordingly, by the evening of 10 May – the same day the German offensive began – British cavalry units had moved up to the Dyle. They were to remain a mere five days. To their north-west the Netherlands collapsed and surrendered in four days. The German attack on the Netherlands was built on a concept not of fighting *through* the country, but of destabilising its defences by means of the early capture of strategic bridges far to the rear by air-landed and parachute forces. This would allow the panzer forces to drive through to carve up the country and so dislocate its defenders. By 13 May panzers were on the outskirts of Rotterdam. A day later, following the devastating aerial bombardment of the city, the Dutch government surrendered. The German plan, strategic in its intent and cleverly executed, was also strategic in its outcome, bringing about the collapse of the Dutch defences and the capitulation of its government. 'Holland surrendered today, after

Operation *Sickle Cut*, 10 May–4 June 1940

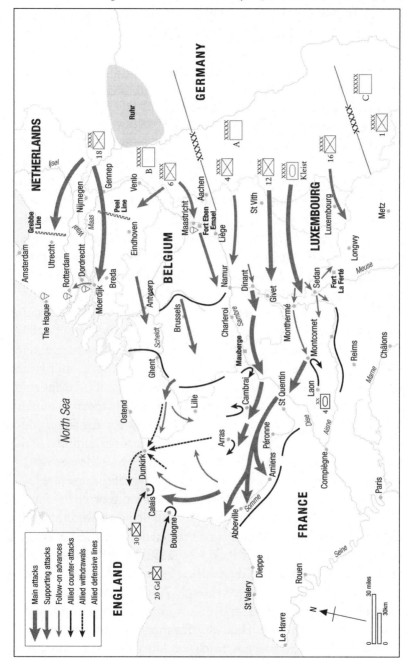

five days of "Blitzkrieg"' recorded Major General Henry Pownall, now Gort's Chief of Staff at the BEF's General Headquarters, in his diary entry for 15 May. 'There was no chance that she would hold for long, but five days is a bit short. The worst effect is likely to be on Belgian morale, already thoroughly bad from top to bottom. They simply are not fighting; long loads of very cheerful troops driving along roads west from Brussels.'[2]

To the south, in Belgium, the fixed Belgian defences on the Albert Canal (the most famous being the fort of Eben Emael south of Maastricht) were taken quickly, those at Eben Emael by 400 combat engineers landing by glider. While for the most part fighting well, the Belgian Army was ruthlessly pushed back, overwhelmed by the weight of force applied to it by von Bock's Army Group B. For several crucial days the pressure placed on the Belgian defenders had the perverse effect on the Allies of persuading them that this was the Germain point of main effort and thus justifying the switch of troops to the north.

However, as most readers know well, while all this was going on, with General Gaston-Henri Billotte's First Army Group staring fixedly on the Dyle, von Rundstedt shocked the Allies by advancing easily through the 'impassable' Ardennes far to the south. The 1,500 panzers of General Paul Kleist's Panzer Group (comprising seven armoured divisions), including General Heinz Guderian's XIX Panzer Corps (800 panzers in three panzer divisions and a motorised rifle regiment), crossed the Meuse near Sedan on 13 May on a 30-mile front The German crossing of the Meuse was not uncontested, and their losses were heavy, but they persevered and pressed on with the knowledge that once this defensive barrier had been cracked, the interior opposition would be considerably lighter. Luftwaffe tactical air strikes against the defenders of the Meuse were significant. French counter-strikes by artillery and air threw back three of the seven attempted crossings. However, the four successful crossings over the Meuse were enough. As the German penetration continued, tactical counter-move decisions by Allied commanders were not necessarily wrong but were invariably late and overtaken by events. The Germans had got inside the Allied decision-action cycle.

Perhaps unsurprisingly, French morale rapidly crumbled, not just on the battlefield but in the Élysée Palace. The new premier, Paul Reynaud (Edouard Daladier had been ousted in March) told the new British prime minister, Winston Churchill, by telephone on 13 May that 'the battle was lost'. Churchill had been prime minister for three days, having replaced Neville Chamberlain in the aftermath of the debacle in Norway. Quite extraordinarily, Churchill had become prime minister as the debate in the House of Commons on the Norway fiasco had become overshadowed by the looming disaster in France. Although somewhat hysterical in tone, Reynaud's comment was accurate, nevertheless. When, on 14 May, the panzers managed to cross the rickety pontoon bridges thrown up by German combat engineers, the French armoured counter-strokes were too slow in both conception and execution to respond.

The German triumph in this campaign comprised not merely speed of movement and the Wehrmacht's ability to combine tank action with artillery fire and air power. A key feature of this attack was that panzer commanders led from the front, sometimes even the leading tank, driving, cajoling, threatening, pressing on with a relentless determination to outpace the enemy by both *thinking* and *doing*. The risks in retrospect were extraordinary. The armoured punch from Sedan over the Meuse didn't extend wider than 30 miles at its greatest extent and stretched, after three weeks, for over 200 miles. It was immensely vulnerable to counter-attack. In 1940, however, apart from a limited British armoured counter-attack at Arras on 21 May, at no time were the Allied armies able to land a blow anywhere near the German armoured vanguard, or eat into its slow tail. It was the poor quality of its French opponents in 1940 which gave *Blitzkrieg* its success. In time, when the Allies knew how to counter such tactics, armoured attacks of this kind, as at El Alamein in late 1942, ran into the sand. In recognition of von Rundstedt's success Hitler, in Directive No 11 on 14 May 1940, acknowledged that the main effort was now to be made by Army Group A. 'Bad news from down south,' wrote Pownall in his diary on 14 May. 'The Germans, inexplicably, have got across the Meuse in the neighbourhood of Sedan and Mézières, a big hole twelve miles wide and ten deep, or more. A counter-attack by the French staged for 11am was postponed till 11.30am and meanwhile the Germans attacked again with more success.' Two days later he wrote: 'The news from the far south is very bad. German mechanized columns are getting deep into France towards Laon and St Quentin. I hope to God the French have some means of stopping them and closing the gap or we are *bust*.'[3]

The contrast between the tactical use of air power by the Luftwaffe and the early response of the Royal Air Force, which remained locked into ideas of fighting a strategic *bomber* battle, while ignoring the armies on the ground, was significant. The RAF contribution to the direct support of the BEF amounted to two squadrons of Fairey Battle light bombers, six squadrons of Lysanders for reconnaissance, and four squadrons of Hurricanes for air protection.[*] It was a half-hearted and wholly inadequate effort, made worse by the realisation during the second phase of the battle, the withdrawal to Dunkirk, of just how damaging air attacks could be against advancing enemy formations. Many German accounts recorded the devastation caused by attacks by light bombers but not enough aircraft had been assigned to co-operation with the ground forces.[†] In contrast to the feeble British air effort, the Luftwaffe's *tactical* combat air capability availability to both von Bock in the north and von Rundstedt in the south was staggering: a total of 2,708 bombers, dive-bombers and fighters (without counting reconnaissance and transport aircraft), all tasked with providing intimate support to the ground offensive. It was a doctrine that was to pay the Wehrmacht enormous dividends in 1940.

In Britain, no effort had been made between the wars to consider the potential for combined air–land operations of the kind the

---

[*]Gort had originally asked for seven bomber and five fighter squadrons.
[†]Robert Kershaw, *Dunkirchen: The German View of Dunkirk* (Oxford: Osprey, 2022). The Royal Air Force in France also had the Advanced Air Striking Force of ten bomber and two fighter squadrons, part of Bomber Command and not available for fighting the land battle.

Germans were to demonstrate in their masterclass of 1940. Indeed, a characteristic of army and RAF relations for much of this time was RAF hostility to the idea that it had a significant role in fighting the land battle, a task that air theorists dismissed as one for the artillery. Right up until the German attack the Air Ministry fought trenchantly to defend the concept of the Royal Air Force being used primarily for fighting the strategic battle against the Ruhr, not in countering the land battle in Belgium or France. The view of the Air Ministry was that a direct commitment of air power to the land battle was 'a prostitution of the Air Force'.[4] Hastings Ismay noted despondently that it 'almost seemed as though the Air Staff would prefer to have their forces under Beelzebub rather than anyone connected with the Army'.[5] Pownall noted in his diary on 11 May:

> It appears that Dill (then Vice Chief of the Imperial General Staff), has lost the battle with the Air Ministry (and Winston Churchill) over the use of heavy bombers. They were to have attacked the Ruhr last night. The C.A.S. [Chief of the Air Staff Sir Cyril Newall] has said that the heavy bombers wouldn't achieve the proposed object if they tried to stop the German advance. If they bombed tonight, Aachen, Maastricht, München-Gladbach even, they would give us some help. Not a bit, they still hanker for the Ruhr while all the time the Germans are pouring in through the gap they've made at Maastricht. This is exactly the sort of situation we foresaw ... the Air Ministry Staff are in process of letting us down completely, as we knew they were planning to do.[6]

When tactical air support was provided to the BEF it was first rate, as Pownall noted two days after the start of the German offensive:

> Our R.A.F. in France have done splendidly, especially the fighter squadrons of the Air Component who are fairly knocking the German Heinkels out of the sky. But our six squadrons are reduced to fifty machines and the personnel has been going all out for three long days. We have asked for four more squadrons

and can get so far only one. That will leave thirty-four squadrons at home where there is no attack. And here the decisive battle is being fought. But the Air Staff refuse to part with any more even 'on loan'.[7]

Aware now that the move into Belgium, thus denuding their southern flank of adequate defences against Army Group A, had been a strategic blunder, Billotte's army group (including the BEF) began withdrawing from the Dyle on the evening of 16 May, establishing a defensive line west of Brussels along the River Scheldt, with an intermediate line along the Dendre. To give Gort his due, he recognised as early as 16 May that the battle was probably lost, and his primary duty would be to rescue the BEF. This isn't how history has remembered him, which is unfair. So much of what he now attempted to do was dependent upon the actions of both French and Belgian allies, as well as a calm appreciation of what the Wehrmacht was attempting to achieve, namely the entire destruction of Billotte's 1st Army Group. With the Belgians on his left flank giving way to the inexorable push of von Bock's legions, Gort attempted to counter-attack where he could, but the cold reality of encirclement from both north and south stared him in the face. London remained out of touch with the speed of events throughout the campaign. Major General Noel Mason-MacFarlane, the Director of Military Intelligence at General Headquarters BEF, let down his guard on 15 May when, briefing the press, he expressed the hard reality: 'History provides many examples of a British Army being asked to operate under appalling handicaps by the politicians responsible for British policy,' he observed, 'but I doubted that the British Army had ever found itself in a graver position than that in which the governments of the last 20 years had placed it.'[8]

On 20 May, with the three leading German panzer divisions racing towards the Channel, Gort was ordered, in instructions brought to him personally by the CIGS, Ironside, to attack the Germans by thrusting south of Amiens. Gort, much closer to the reality of a fast-deteriorating situation, and now determined that his primary task was to preserve the BEF rather than to expend it in fruitless gestures against the inevitable, ignored them. It was at this point that a famous though short-lived engagement took place between a small force of British tanks and a man whose name would subsequently become one of the most famous of the entire war: Erwin Rommel. Rommel's 7th Panzer Division had become unmanageably extended in the days leading up to 20 May, and he was forced to reorganise himself. A counter-attack by two battalions of the 50th (Tyne and Tees) Division and 74 tanks, 58 of which were Vickers light tanks armed only with machine guns and the remainder of the new Matilda Mark IIs, struck Rommel as he was attempting to move around Arras. For the first time in the ten days of fighting so far in the campaign a running tank-on-tank battle took place. The British tanks, however, were slow, and the advance into the German columns, while causing some confusion to start with, was met by a hurriedly organised anti-tank defence. Rommel personally organised the defence of his columns, using his 88mm flak guns and field artillery in the direct fire role, in time chewing up what had been a spirited British attack. The desperate fact was that the British tanks had advanced into battle entirely untrained in any doctrine of the armoured offensive, or in the concept of counter-penetration. These brave men had been thrown into battle in new vehicles that were poorly designed for the type of fighting they would encounter, and untrained in the art of armoured manoeuvre. The attack worked for a time, at least against thin-skinned enemy vehicles and unprotected infantry, but ran out of steam when countered by a fierce defence overseen personally by the German divisional commander himself.

Astonishingly, the tanks of Panzer Group Kleist were now threatening the Channel coast at Abbeville, near the mouth of the Somme in an unbelievable – certainly to those German veterans of the Great War who found themselves fighting in France again – three weeks. 'Nothing but a miracle can save the BEF now,' wrote Brooke in his diary.[9] On 25 May Gort unilaterally decided not to support his instructions from London for a combined counter-attack to his front, because he wasn't confident that it could be carried out and, fearful of the security of his left flank, he made the decision that saved the BEF from annihilation. He withdrew the 5th and 50th Divisions under Brooke's II Corps and moved them to cover his left flank, protecting his force from the imminent collapse of the Belgian Army, which surrendered on 27 May. Gort's move was made in the nick of time, Brooke's corps arriving in their positions just in time to halt German attacks on this vulnerable flank.

From this point on Gort had one mission, an implied task that he had worked out for himself: to protect and evacuate the BEF, to allow it to fight another day. This, in late May, was a distant prospect. On 26 May Gort was ordered to try to rescue the BEF by withdrawing it for evacuation along the Dunkirk beaches. Ironside, the CIGS, wrote in his diary on 27 May that he feared that only 30,000 men would be evacuated from France.[10] In the event, of the 394,165 in the BEF in April 1940, 'only' 68,111 were lost: killed, wounded or taken prisoner of war.* Without Gort's precipitate action on 25 May to redeploy Brooke's corps to his northern flank, there would have been no Dunkirk. Major General Bernard Montgomery, commander of the 3rd Infantry Division, agreed. 'Gort saw clearly that he must, at the least, get the men of the B.E.F. back to England with their personal weapons,' he wrote in his memoirs. 'For this I give him full marks and I hope history will do the same. He saved the men of the B.E.F.'[11]

*Of this number, 40,000 became prisoners of war.

Chapter 17

Sichelschnitt: the anatomy of disaster

The world has largely remembered *Sichelschnitt* as a brilliant German operation of war, but it was one that was fundamentally enabled by Allied ineptitude. Indeed, *Blitzkrieg* wasn't particularly new, innovative or even a warfighting doctrine. It is best described, in the context of France 1940, as an *event*. It was simply the way that the Wehrmacht exerted its tactical and operational superiority over its more pedestrian enemies in 1940. In fact, it was the 1940 extension of what the German Army had first demonstrated in Flanders in March 1918, this time with tanks and Stukas. It was the *Panzerwaffe* ('tank force') – combined with a tactical air force – which in 1940 would create the breakthrough that Ludendorff had been unable to achieve in 1918. Where it was applied, by Army Group A, it concentrated fast-moving armoured vanguards co-ordinated with tactical air power, such as 400 Ju 87 Stuka dive-bombers, to so overwhelm the enemy in both time and space that they were unable to respond quickly enough to the changing and challenging battlefield. In 1940 the panzer came into its own, the sound of clattering tracks on French cobblestones a new feature in the sound of battle and a key element in how France remembers its defeat in 1940.

It wasn't the type of tank in the German inventory which mattered, but the way in which these tanks were employed. Only about 10 per cent of the army comprised tanks, the remainder relying on horse and wagons and the raw, painful feet of the marching infantry. Of the 2,539 tanks the Wehrmacht deployed in 1940, only 916, or 36 per cent were battleworthy, the remainder being clattering tin cans with machine guns (the obsolete Panzer Mark Is and Mark IIs). The only modern tanks were 683 Panzer Mark IIIs and Czech T38 tanks armed with a 37mm gun, and 278 of the larger Mark IVs with a short 75mm gun. But it was enough. The German operational strategy was to use this mass of armour not to fight a large confrontational tank battle, but to achieve breakthrough and breakout, bursting through the enemy's linear defences. It was surprise and shock action that so discomforted the Allies, who had lazily and, given what we know of British failure to understand 1918, ignorantly assumed that the war would progress against a 1914 rather than a 1918 pattern. The armoured vanguard would surge through the outer skin of the enemy defences, concentrating heavy effort in one place, before driving hard into the heart of enemy territory. With an enemy intent on fighting a linear battle, the rear areas, behind this outer crust, would be weakly defended and full of rear-echelon, administrative and supply troops managing the lines of communication up to the front, not expecting to have to fight. It was by driving hard and fast behind the enemy front line, breaking the cycle of Allied battlefield decision-making, that *Blitzkrieg* was to achieve its psychological effect.

In contrast the Allies remained concerned about retaining the integrity of their defensive lines. The diaries of Major General Henry Pownall, for instance, are replete with concerns as the days spun past about the widening frontages on one defensive line or another. British concern was misplaced. It was to spread the ever-decreasing butter of the British infantry across ever-widening stretches of French and Belgian bread, without realising that the Germans were concerned not with rolling up a front line, but with driving hard to the rear. By so doing they would take risks with their flanks, but the discombobulatory effect on the enemy was considered to far outweigh any worry about the risk of counter-attack from an increasingly battered and disorganised enemy. Of course this operational concept was risky, but the risks taken were carefully calculated given what the German General Staff knew about British and French tactical doctrine, or the lack of it.

These German tactics were psychologically disconcerting for those not trained to expect them. As was demonstrated on the Meuse, artillery would batter a position in co-ordination with armoured columns bypassing fixed defences and attacking those it needed to clear from the flanks and the rear. The infantry accompanying the advancing armour – *Panzergrenadiers* (mechanised infantry) – arrived in tandem with the Stukas, which could drop their bombs from a screaming dive. Each Stuka seemed to those at the receiving end to be diving directly at them, personally. For untrained troops it was a terrifying experience. The panzers would sweep on while the truck-borne infantry would turn up to deal with survivors of this storm of fire and movement. By this time, of course, the disorientated French and British would now consider themselves cut off, behind their front line, with no prospect of being relieved. Surrender or a disorganised escape to the rear would seem to be a more sensible option than the forlorn hope of continued resistance when the surrounding fields were dotted with the grey-green uniforms and coal-scuttle helmets of their enemy. The psychological effect of *Blitzkrieg* was considerable. This wasn't how their fathers had told them war was fought. How did the Germans manage to discomfort them on the battlefield so comprehensively? Were they inadequate soldiers, unable to meet the standards of campaigning set by the previous generation? Or was it that their tactics were simply not able to cope with the shock of a comprehensive assault by German infantry, armour and air power all descending on them at once? This was the battlefield that the British had entirely dominated, by virtue of their tactical innovations, in 1918. It was now Germany's turn, a direct result of the failure of the British Army to develop its doctrine and approaches to warfighting at the end of the Great War. Brave men in 1940 did their duty, but against a battle-winning concept of their enemy, they were out-thought rather than out-fought. And critically, when an army thinks it is beaten, it is indeed beaten.

Sichelschnitt was thus not what Britain or France had expected. Nor, for that matter, was it what the Wehrmacht had originally planned. A coterie of smart-thinking officers in von Rundstedt's army group, led by his chief of staff Eric von Manstein, saw an opportunity to completely destabilise the entire Allied plan. Manstein's plan was not initially to race to the English Channel, but to surround and destroy the French and British armies in north-western France in what the German doctrine described as a 'cauldron' battle: *Kesselschlacht.* It was a standard operational-level German approach to the offensive, demonstrated at the Battle of Tannenberg in August 1914 when the Russian General Samsonov's Second Army was encircled and annihilated by the German Eighth Army. Such dramatic operations would subsequently characterise the early stages of Operation *Barbarossa* in the Soviet Union from June 1941. Deep infiltration – against identifiable weak points in the enemy defence – using fast-moving panzers would be the vehicle by which the enemy front would be penetrated.

What wasn't expected in France was just how easy it would be. Within three weeks of crossing the start line Guderian's exhausted though jubilant panzer troops – with the walking infantry and horsed supplies far to the rear – were looking in awe at the English Channel, a triumph of which their forefathers could only dream. They failed to execute a *Kesselschlacht* by virtue of the Allied withdrawal and the evacuation of the BEF from the beaches of Dunkirk, but they had nevertheless achieved an extraordinary success. The warfighting capability of the army that Britain had rushed to raise and equip after March 1939 was destroyed in France and its equipment left at Dunkirk – 12,200 artillery pieces, 1,350 anti-aircraft and anti-tank guns, 6,400 anti-tank rifles, 11,000 machine guns, 75,000 vehicles and virtually all its scarce supply of tanks. The Germans turned south and Marshal Pétain[*], humiliated, offered a ceasefire and on 22 June accepted German terms.

[*]Premier of Vichy France.

Dunkirk certainly represented profound military failure by Britain and France at both the strategic and operational levels of war. At the tactical level, most accounts show that the soldiers of the BEF, poorly prepared, nevertheless gave a good account of themselves. During the course of the campaign no British formation was forced out of its defences or over-run. Every withdrawal which took place was carried out under orders, to conform with the general plan. But wars are not won by good tactics or even by skilful fighting: they are won by using the right operational strategies. Equally, as Britain and France demonstrated in 1940, they are lost by poor operational plans and mistaken grand strategies. Disaster for Britain was only averted by the discipline and courage of the British soldier. Out-thought, out-manoeuvred and ultimately out-fought, the morale of the British Tommy did not break, leaving hope for another day.

In Calais, struck by waves of Stukas, artillery and finally an infantry assault, the depleted 60th Rifles fought on until dead or wounded. Gris Davies-Scourfield's end came when, running through some back gardens to get to one of his platoon houses, he was shot:

> One bullet went through my right arm and another hit me in the side ... it felt as if I had been clouted in the ribs with a sledge-hammer ... I staggered a few steps towards a burnt out truck, hoping to find cover behind it, but I was hit again, this time in the head. I was conscious of starting to fall and then must have passed out.[1]

Later, when it was dark, a German soldier found him, put a field dressing on his head wound, and saved his life. For Lieutenant Davies-Scourfield, the war was over. He was to spend the remainder of the war as a prisoner of war. In the smoking ruins of Calais, and along the sand of the beaches of the Pas de Calais lay the detritus of a smashed and humiliated army. A mere 22 years on from the defeat of the Central Powers in 1918, disaster enveloped the second

British Expeditionary Force in France. The German *Blitzkrieg* swept all before it and what was left was forced into a series of humiliating evacuations from the beaches and ports of northern France. Davies-Scourfield was one of 68,111 British casualties. The gallant defence of Calais could not quite hide the fact that lives had been sacrificed for a failure deep rooted in pre-war British polity and military planning.

General Sir David Fraser, author of a history of the British Army that tracked its early defeats in the Second World War to eventual victory, summarised in his memoirs the military reasons for the defeat of the BEF in the summer of 1940, a cataclysm that led to the sands of Dunkirk.

> ... we, the French and British, had been beaten not simply, not even primarily, by superior or more modern equipment, although the enemy had outmatched us in mechanised vehicles, in the air and in the emphasis, he had clearly put on using air power as an adjunct and close support to land operations. Nor had we been beaten by superior numbers. *We had been beaten by dynamism; by entirely traditional 'grand tactics' or operational doctrine;* by excellent minor tactics, fitness and discipline; but above all by an army imbued with high, confident morale and the spirit of the offensive. We didn't appreciate such things at the time, but – particularly at the beginning of a war or a campaign – he who attacks has a priceless initiative. He can plan, rehearse, choose his place and his moment. For him who defends there is a two-fold penalty; he does not know what to expect because the whole experience of war is new. *By their absolute determination to stand on the operational defensive (drawing false lessons from partial perceptions of the previous war) the Western Allies made their defeat virtually certain.*"[2]

---

*The italics for emphasis are ours.

It is easy to blame the army for the events in France that led up to Dunkirk in June 1940. Indeed, in assessing the events of the six weeks of May and June 1940 which led to the destruction of the BEF and the French 1st Army Group, most analysis has tended to focus on the *battle* for northern France, especially the quite extraordinary masterstroke by von Rundstedt's Army Group A, which has come to define, in British minds at least, the brilliance of *Blitzkrieg*. Von Rundstedt's scythe through the Ardennes, the dramatic armoured dash from Sedan to the Channel coast and the resulting embarrassment of defeat for both Billotte's 1st Army Group and the BEF (of which it was a part), have all the high drama that made Hitler believe himself to be invincible. After all, the evidence seemed to demonstrate it: Poland had been conquered in four weeks, Denmark and Norway in eight, the Netherlands in four days, Belgium in seventeen and France in six weeks. It has also become the defining story, with the exception of the Battle of Britain and the London Blitz of 1940. The year was truly the *annus horribilis* to beat all others in Britain's history. The story of the land campaign in France and Belgium in 1940 has tended to be one that stresses the superiority of German operational doctrine, characterised by tanks operating in fast-moving armoured vanguards in which German commanders dictated the pace of the advance, thinking faster and acting more quickly than their opponents, and allowing better-deployed armour to do the talking. It was Allied strategic failure, however, that led to the debacle in France in May and June 1940 as much as it was German operational success, for without the first, the second could not have been possible.

The British Army failed to measure up to the German *Blitzkrieg* in 1940 because it couldn't get its head around – let alone respond effectively to – the tactics deployed by von Rundstedt's Army Group A. Nowhere in any of the appreciations of German operational objectives was it considered that the Wehrmacht might attempt to punch through the outer crust of Allied defences with the aim of creating the conditions for a cauldron battle. Because it couldn't comprehend the type of war it would have to fight, it came to the Second World War entirely and grievously unprepared. It did not understand the nature of *Blitzkrieg*, because it had failed to learn the lessons of 1918 and develop for itself during the inter-war years a doctrine of fast-moving combined-arms manoeuvre, able to achieve strategic as well as tactical objectives. (Nor too had France, for what it is worth, which had a far larger and arguably stronger army.)

Had Britain done much more during what Churchill lamented had been 'the years that the locust had eaten'* to understand and prepare for the next war, it would have understood the power of tank-led, combined-arms formations to disrupt and confuse enemy decision-makers and therefore the need to defend against them. There is much in the argument that the new German Wehrmacht developed these operational strategies because it was the underdog and it needed to win quick victories. Shock and surprise were key elements in this approach. Whatever the case, German strengths in 1940 lay in both operational manoeuvre and tactical expertise. The Reichswehr had analysed 1918 far more thoroughly than the British. It had seen the near-success of Ludendorff's March 1918 offensive, and the much greater and sustained Allied success of the Hundred Days. During the 1920s and 1930s it created a doctrine of armoured formation manoeuvre when it had no tanks, and few men. It had assiduously followed technical and doctrinal developments in other countries, including the Soviet Union, France and Britain. It had developed its own tanks, for much of the 1930s in partnership with the Soviet Army, far from the prying eyes of the Versailles policemen. It thought, drilled and trained relentlessly for the next war which, to Germans of all stamps, was increasingly an inevitability.

Britain, by contrast, had not. The result was that, when initially deployed to France in 1939, the BEF meekly accepted a French strategic plan that involved the establishment of lines of fixed

---

*Joel 2:25. This phrase was used by Winston Churchill to the House of Commons in November 1936 to describe the years lost to Britain's security as a result of Baldwin's and Chamberlain's policy of appeasement. If he was being honest, he would have acknowledged that his own policies as chancellor in the early 1920s had been a significant cause of the problem.

defences ('trenches') across northern France that were designed to be an extension of the Maginot Line, together with a plan to thrust troops into now neutral Belgium to resist a German attack, without considering the strategic implications of both ideas. If British commanders had studied and trained in the concepts of combined-arms manoeuvre, they would have recognised the extreme vulnerability of the positions they were trying to establish. The Hindenburg Line hadn't helped the Germans in 1918; anything similar (and the British trench lines in 1940 were not a patch on the Hindenburg Line) wouldn't survive in 1940. As the more imaginative German commanders such as von Manstein, von Rundstedt and Guderian knew, combined-arms manoeuvre gave military commanders choice, speed and the ability to switch their armoured thrusts at will. In any case, wedded to a Maginot Line-type concept of defence that assumed that only linear defences protected territory, the French High Command would have rejected any thought of turning north-western France into a model of defence in depth. While this would have been the only thing to stop von Rundstedt's armoured columns it would at the same time have traded precious sovereign French soil, which was anathema to both French soldiers and politicians. 'They Shall Not Pass' had been the French Army's *cri de coeur* in the Great War. Anything that did the opposite this time around would never be acceptable. The BEF was therefore beaten the moment the Germans crossed into France.

There were two possible counters to the threat of armoured penetration or infiltration that the Germans were to deploy in 1940. The first was the idea of mobile counter-penetration. These are combined-arms tactics designed to deploy strong anti-armour forces anywhere on the battlefield quickly, to destroy (using mines, infantry and artillery-crewed anti-tank weapons, tanks and tactical air power) enemy armoured columns and their supply lines. The British Army was light years from having this type of tactical doctrine in 1940, and arguably didn't develop a doctrine of counter-penetration until the 1980s, when it became a key strategy for halting the operational manoeuvre groups of the Soviet Third Shock Army.

The only other defensive counter to this threat was fixed defences in great depth all the way back to the Channel coast, powerfully defended bastions (or 'defensive areas') connected by plentiful artillery and tactical firepower purposed to draw in the attacker, disperse him and defeat his infiltrating columns piece by piece. It would have meant sacrificing ground (it being far less important than the objective of forcing the enemy's forces to be whittled away); it would have required experienced and well-trained troops, willing to be temporarily isolated and cut off as the enemy armies swarmed around them until they could be dispersed and destroyed. It would also have needed large quantities of heavy mobile artillery; considerable amounts of tactical air power able to be called on by bastion commanders to counter-attack enemy infiltration and considerable quantities of anti-aircraft artillery (or counter-air assets such as fighters) able to take the sting from the widespread use of Stuka dive-bombers. In short, it would have required a modern army trained, equipped and able to fight as it had done in 1918, now with modern tanks, aircraft, artillery and other equipment. It would have needed commanders trained over years of peacetime field exercises and table-top exercises to design and deploy a counter-*Blitzkrieg* defensive strategy in northern France that could negate the power of a combined-arms offensive that bypassed the fixed defences of the Maginot Line. On this point it is worth noting that Hitler probably learned the wrong lessons from *Sichelschnitt*, believing that he had discovered the Holy Grail to the successful offensive. The Eastern Front was to demonstrate in time that such offensives could be countered: all that was needed was tactics of counter-penetration (deployed subsequently by both the Germans and the Soviets) able to defeat in detail the enemy armoured vanguards, using tactical air power, mines, anti-tank guns and tanks.

The fact is, however, that in 1940 neither Britain nor France had a military solution to the deep penetration undertaken by Army Group A. It is hardly surprising that at an operational and a tactical level the BEF were surprised by the speed with which Army Group C operated, and by the fast-moving German columns, for instance, accompanied by tactical air power (the infamous Stukas) which characterised the race to the sea. They were not tactics with which the British Army was familiar. This was the direct product of two decades in which the British Army had paid no attention to the complexities of operational-level warfighting (either offensive or defensive) in Europe. The possibilities for fast-moving, shock-inducing armoured infiltration, of the sort that Army Group A delivered during those extraordinary summer weeks, had never been considered, let alone been wargamed on a table or in a sandpit.

In 1940 the brutal lesson taught to the French and British was the need to remain current with the theory and practice of war. Because the British Army did not operate at this higher operational level in 1940 – and arguably did not until late 1942 – the Wehrmacht possessed a temporary advantage in 1940. Not preparing for a peer-on-peer conflict of the type that had engulfed the BEF in 1940 demonstrated that the army was very far indeed from being considered an equal to the German Army when it mattered. This was a profound failure, the direct responsibility of successive governments in London who had assumed naively for two decades that nothing like this could happen again in Europe, at least not in the short or medium term. The events of May 1940 made such assumptions look foolish in retrospect.

This operational deficiency was compounded by the strategic blunder of moving troops with no preparation into Belgium at the outset of the fighting. The blame for Britain's military humiliations in France 1940 lay not with its benighted troops in Europe, operating to a defensive strategy that took no account of the potentiality for Wehrmacht operational brilliance of the sort seen in *Sichelschnitt*, but with the series of governmental decisions going back to the country's dismantling of its military capability after 1918, the worst of which was the dreamland notion that Britain could get away with a policy of 'limited liability' in Europe. Not wishing to get involved in war on the Continent was never enough to prevent it from happening. Not only did Britain do nothing to prevent the Second World War by using its armed forces to threaten retaliation against German remilitarisation (e.g. the Rhineland in 1936) or aggression in Europe, but by its political naivety and political ineptitude between 1933 and 1936 it enabled war to become more, rather than less, likely. Politically and diplomatically Britain during the 1930s was scared of its own shadow. The consequence of this profound failure of self-confidence was not just defeat in battle in 1940 but the failure to prevent war in the first place. The critics of *si vis pacem, para bellum* had been proven grievously wrong.

For France, May 1940 was nothing short of a catastrophe, as it entailed not just the defeat of its army, but the loss of its country and liberty. The strategic effect of the attack by Army Group A was not tactical, or even operational. It was the strategic disembowelling of France's martial pretensions, a grand humiliation that served as a psychological blow to the idea that France was mature and sophisticated enough to defend itself against a determined and energetic foe. *Blitzkrieg* showed, to the French, the failure not just of France's army and its pre-war defence planning, but of the French character. Unlike their parents' generation, it was felt, the French no longer knew how to fight. It was a mental body blow to the French psyche, the echoes of which continue to sound down the years.

For its part, the BEF was lost long before it was deployed to France in September 1939. Only when the British Army could fight on equal terms would it ever hope to be successful in this type of warfighting. No amount of 'Miracle of Dunkirk' sugar coating can change the fact that in both its *thinking* and *doing* the BEF was never capable of countering, let alone defeating, an adversary as formidable as the Wehrmacht in full throttle in the summer of 1940. Defeat in battle was the direct result of the egregious failure over the previous two decades to prepare adequately for the kind of warfighting that many suggested loudly would come, but which was comprehensively ignored by the British government, the British people and the British military establishment until it was too late. All must take a share of the blame for the savage defeat suffered by British arms in France in 1940.

Soon after the return of the BEF to British shores a committee of seven was established under General Sir William Bartholomew, previously of Northern Command, to investigate the lessons of France for the organisation and training of the British Army. The committee's final report made a number of pertinent observations and recommendations, most of which were concerned with tactical changes and enhancements that the army urgently needed following the sudden shock of war in France.[3] 'New and unexpected developments included the dive-bombing attack, and a greater number of enemy armoured formations than had been anticipated,' the report ran. The committee recognised that the operational art deployed in particular by Army Group A was psychological, designed to disorientate the enemy so that the cauldron battle could be decisively enacted. There is no evidence, however, that the committee recognised this German operational goal in 1940. It appreciated the success of Gort in bringing the BEF back to a point by which it could be evacuated to Britain but did not understand that by withdrawing Gort had rescued the BEF from the cauldron. But the focus of the assault on disorientation rather than hard battle, at least in the encircling phase of the cauldron operation, was very clear. The committee observed that 'By every means in his power, and often with great ingenuity, the enemy has concentrated his means of attack on the morale of his opponents. In the application of his weapons, he relies almost as much on "terrorization by noise" as on material effect.'

The committee suggested that both the brigade and division have built into them artillery, anti-aircraft, medium machine-gun, armoured reconnaissance and anti-armour units, a revolutionary argument for the British Army, which had so far shied away from this level of integration. It shied away, however, from creating a truly all-arms formation by incorporating within it an armoured unit or units, although it advocated a much greater independence of action for its brigades, which it suggested should be equipped to fight independent actions when necessary. This was an argument that was to vex the British Army for some time thereafter: what was the level of organisation at which full integration should be achieved? It all depended, so the argument went, on what formation would fight; was it by brigade or by division? If brigades were to fight independent actions, they would require a corresponding complement of armour, anti-tank weapons, artillery, combat engineers and so on.

The whole tenor of the report was about improving the quality of the defence: the observations it made about the offensive tended to relate to how to counter the tactics deployed by the Germans, rather than any fundamental reconsideration about how British tactics might need to change to enable the army to exploit the power of manoeuvre. One finding was that German operational manoeuvre (punching through weakly defended areas along the Meuse with armour) was undertaken in conjunction with concentrated firepower – that of artillery and tactical air power (primarily Stukas). Troop-carrying aircraft, enjoying considerable measures of air superiority, even brought up additional infantry. Air superiority also prevented the BEF from using its own aircraft for artillery spotting.

> The German method of preparation for attack consists of rapid reconnaissance, which taps along the front line until a weak spot or gap is found. As soon as such a spot is located the crossing of the obstacle is effected and a small bridgehead made. This bridgehead is subsequently widened, and a tank bridge speedily installed over which the tanks cross. Once a crossing is made the bridgehead is widened to allow the passage of more and more troops. If a weak spot is not found as a result of the initial reconnaissance, a concentration of gun and mortar fire, or dive bombing is put down behind which a crossing is forced. The ensuing action is in both cases the same.

Subsequent to the forming of the bridgeheads, which may happen at one or more places, the Germans, by using infiltration methods and with a complete disregard for open flanks, attempt to push their mobile troops through, if necessary, on a narrow front. The axis of such an advance will sooner or later be one of the roads leading back through our position.

The Bartholomew Report was a calm, rational and detailed analysis of the tactical problems that had presented themselves to the BEF in the sudden clamour of battle in France. It represented for the British Army, however, something of the pain of the dentist's drill – without anaesthetic – after years of decay. Recommendations were made about tactics to deal with massed tank attack (mines, tank-hunting infantry units and lots of 6-pounder anti-tank guns); the superb use of tactical air power by the Germans which the British urgently needed to replicate; the need for fixed and mobile anti-tank defences to be 'in great depth'; the need for defensive positions to be primarily offensive rather than static or sedentary; the need for all troops, when surrounded, to realise that the enemy passing them will soon exhaust their supplies, and therefore to stay put, and not panic about the enemy being to their rear; and the need for surprise at every stage of the battle.

There were at least two problems with the Bartholomew Report. In the first place, it was so concerned with identifying German approaches to battle and filling gaps in British defensive armaments that it made virtually no recommendations about how British offensive tactics needed to develop, though there were some helpful observations about how battalions, brigades and divisions needed to be reorganised to reflect the need for much more combined-arms unity on the battlefield. These observations were akin to calling for infantry battalions and brigades to become something like combat teams, possessing inherently a range of combat capabilities that mirrored what the BEF had seen the Wehrmacht deploy in battle. But, reflecting the experience in France, the exclusive concern was defence, an understandable preoccupation given the very real fear in 1940 that the Germans might soon be launching an invasion.

The time would come nevertheless when the British Army would deploy its fighting divisions to attack. How should they fight?

The second problem with the Bartholomew Report is that it sank quickly into obscurity amidst the immense task the country faced in reconstituting the army and preparing it for future battle. As was the situation in 1939 when the army was preoccupied with a rapid and unexpected expansion to create an expeditionary force for France, mid-1940 saw the army attempting to reconstitute itself from the detritus of defeat. It had few weapons, for instance: most had been left on the beaches of Dunkirk. There was no time in the circumstances to contemplate the complexities of formation tactics, especially those that represented anything more than simply attempting to organise masses of infantry to dig in and hold a line. That would have to come later when Britain had an army again. The problem was compounded by the longstanding problem that there remained no mechanism for capturing development ideas and converting them into new training and new doctrine.

Following Dunkirk, the British Army continued to stumble forward, in terms of its approach to the construction, exposition and execution of its military doctrine. There remained still no uniform answer to this most fundamental of questions: what was the most effective way of organising itself for battle?

Epilogue

El Alamein and beyond

After the humiliation of 1940, the British Army struggled to identify a means of defeating the operational mobility of the Wehrmacht. The Western Desert came to be a literal sandpit for competing theories as to how the one general who came to be regarded as the arch-embodiment of German tactics – Erwin Rommel – could be defeated. In 1941, with Rommel's arrival in Africa, a doctrine-deficient British Army struggled to understand – as it had in France in 1940 – German operational manoeuvre. It was ironic that in the Western Desert in 1941 what had been dismissed before the war as the theoretical extravagances of the 'prophets of mobility' were seized upon as attempts were made to counter German tactics. The problem during British attempts to raise the siege of Tobruk in late 1941 (Operation *Crusader*) was to assume that German armour operated like a battle fleet at sea and that all that was required was to lure it into a decisive battle. It did nothing of the kind. As the historian Brian Holden Reid explains:

> In fact, the German armour did not fight like naval fleets, but as the armoured theorists had recommended, in close co-ordination with their anti-tank guns in an integrated defensive-offensive. This was overlooked by Eighth Army staffs, who, seduced by impressive sounding tags, failed to study the substance of their meaning. Under the pressure of war itself it is perhaps not

surprising that they committed this sin. But whatever the merits of the staff or their commander, it must surely be agreed that if the British Army had had recourse to an accepted body of doctrine that Cunningham's plan could have been based on it and that it would have served as a focus for his judgement, experience and intuition, even though he had come from East Africa, where he felt at home. Cunningham would not have had to start from scratch which he found so emotionally and mentally exhausting before the operations began.[1]

An absence of adequate doctrine meant that ill-prepared British commanders – in this case Lieutenant General Sir Alan Cunningham, commander of the Eighth Army – struggling to find an answer to German tactics, were forced to clutch at straws. One such straw was that the primary fighting formation should be the armoured brigade, because infantry-dominated divisions were too vulnerable to attack by enemy armour. This tactic was undone, however, by the fact that the British did not understand that German tactical success was based on the close working relationship of panzers and anti-tank artillery, a mix of the tactical offensive–defensive. Although the British had an excellent anti-aircraft gun, the 3.7-inch, which could have been used to defeat German armoured advances, they didn't have the tactics to deploy them in a similar way to that in which the Germans deployed their famous 88mm guns. British tank columns would be regularly lured into German anti-tank ambushes. Battle is not the place to introduce radical innovations to an army that was not structured or disposed towards radical change.

A severely logistically challenged Afrika Korps was eventually brought to its knees at El Alamein in late 1942 not by the fancy footwork of brand new British armoured doctrine, but by the old-style methodical, carefully choreographed application of firepower. It was no surprise that the architect of this British victory, the new commander of the Eighth Army, Lieutenant General Bernard Montgomery, applied all the principles of 1917 (rather than 1918) in its application. It worked. If the British Army couldn't out-manoeuvre the Desert Fox, it could bludgeon him. In Monty's view, crushing tactical attrition was the only solution to the operational manoeuvre deployed by his opponent, especially with a largely citizen army that had proven itself unable in the heat of battle to respond quickly to German tank-led operational manoeuvre. With the massive quantities of new American weapons and equipment flooding in to North Africa materiel superiority made Monty's task possible, in a way denied to Auchinleck and Cunningham before him. It was the approach adopted for the remainder of the war in Europe: the carefully co-ordinated, professionally choreographed and overwhelming application of force. Monty's success lay in imposing an approach to warfighting that, in the absence of any other doctrinal templates, enabled Britain's citizen army to use its increasingly dominant materiel strength and far greater firepower to win its battles. It is no surprise that the title of the document explaining the British Army's approach to war in 1985, before the adoption of the Bagnall reforms and thereafter the first of the subsequent Army Doctrine Publications,* was unashamedly entitled 'The Application of Force'.[2] It was in essence a recapitulation of the Principles of War (upon which the Field Service Regulations had been built), replete with examples going back to the Peninsular War. 'The Application of Force', in homage to the master, even had a photograph of Monty on its front cover. This was the standout British approach to battle after late 1942 and for most of the Cold War.

One might have hoped that the British Army, having learned its lesson in 1940, would have embraced the necessity for clarity of thinking about doctrine following the end of the war in 1945. Not a bit of it. The first attempt to formulate an army-wide doctrinal approach to the deployment of the army in conventional (i.e. non-nuclear or counter-insurgency warfare) was not to take place until 1988, just as the Cold War was fast accelerating towards its

---

*Preceded by 'Design for Military Operations', *The British Military Doctrine* (HMSO, 1989).

end. The British Army had entered the Second World War without a single, coherent warfighting doctrine, and then exited it in 1945 without acquiring one on the way. Instead, it had found a sticking plaster in Europe – materiel superiority or attrition ('firepower') – which offered something of a convenient substitute. When, in an attempt to introduce a warfighting doctrine to the British Army in 1989, every officer of captain rank and above was sent a copy of the slim booklet 'British Military Doctrine', arguably 71 years after it was first needed (or at least 44 if we take 1945 as the start point), in the preface Field Marshal Sir John Chapple, who had succeeded Sir Nigel Bagnall as CGS, felt moved to urge officers not to dismiss the slim volume (87 pages) as a new-fangled innovation. 'There may be some who say that laying down doctrine like this is not the British way,' he wrote. 'To them I would say that as a nation we have had no shortage of original military thinkers. Also, the modern battlefield is not a place where we could hope to succeed by muddling through.'

Why did the British Army stick with Monty's approach to war, rather than formulating a new warfighting doctrine, for so long? The answer was that it perfectly suited the circumstances of the time, just as Monty's approach suited the situation in North Africa in November 1942. As Lieutenant General John Kiszely notes:

at the end of the war many veterans and commentators believed that, ultimately, success had been retrieved by Montgomery's highly centralised approach where superior materiel was applied against the enemy's strength in accordance with a rigid master plan.

As CIGS in the immediate post-war years, Montgomery's view, not unnaturally, carried considerable weight, and his admirers occupied influential posts in the army for many years after the war. Staff College instruction at the time, unsurprisingly, emphasized a highly controlled style of warfare which sought to impose order on the battlefield, moving firepower to destroy enemy strengths, rather than a style aimed at flourishing in situations of chaos and uncertainty.

Furthermore, the attritional approach appeared to many to be entirely compatible with the British Army's main role as the British Army of the Rhine. A large proportion of the army settled down in West Germany to confront the Russian, and subsequently Warsaw Pact forces limited by constraints which allowed little room for manoeuvre. A strategy of forward defence for reasons of political necessity within the NATO Alliance resulted in a linear and heavily positional General Defence Plan, with lateral movement constrained by a number of national corps deployed shoulder-to-shoulder in what became known as the 'layer cake' deployment. Formations and units found themselves planning positional battles and a slogging match – how best to absorb the shock of the enemy attack rather than avoid it and requiring defence of ground to the last man. Although substantial elements of the British Army were deployed elsewhere, for example East of Suez, training emphasis at the Staff College and at arms schools was firmly on the type of warfare anticipated in Europe.[3]

The parallels with the linear concepts that spawned the Maginot Line were remarkable insofar as, following all the contrary evidence of 1940, they retained their hold on defensive doctrine in central Europe well into the 1980s. They were only replaced when the concept of operational manoeuvre was propelled into British doctrine by the advocacy of General Sir Nigel Bagnall in the early 1980s. Extraordinary as it might seem, given the lack of such things in the 1930s and 1940s, Bagnall instituted standard operating procedures across the I (British) Corps, common to all units and formations for the first time. The task of reducing a Soviet offensive into Western Europe, this doctrine asserted, was best achieved by reducing the power of attacking Soviet columns ('operational manoeuvre groups') by means of manoeuvre itself, rather than through fixed lines and layers of defences that were largely immobile when battle was joined. It was akin to arguing that a von Rundstedt or Rommel thrust was best countered not by fixed defences but by defence in depth and by means of a von Rundstedt or Rommel-type *counter*-thrust. This became, in due course, part of the 'manoeuvrist approach' to warfighting as a foundation stone of British doctrine. Instead of bludgeoning him to death, commanders were now encouraged to seek to defeat the enemy by shattering his moral and physical cohesion – his ability to fight as an effective, co-ordinated whole – rather than necessarily to destroy him physically through incremental attrition.

In 1918 the British Army, after four long years of war, and through trial, perseverance and the application of considerable thought and experimentation, not to mention blood and treasure, had found the formula for success in modern, industrialised warfare. This was decisively demonstrated in the dramatic success of the Hundred Days campaign between August and November, which destroyed the German armies in Flanders and brought an end to four years of terrible fighting, the early years of which had evidenced insufficient imagination or military innovation. Indeed, it was imagination, initiative and innovation – on top of manpower and material superiority – that gave the Allies a war-winning advantage in the final quarter of 1918.

As this book has shown, however, in the years after 1918 the British Army failed to build on its battlefield success in the Hundred Days to create a coherent warfighting doctrine. It was a core understanding of *how* it would fight a high-intensity war that was the essential foundation for driving future capability development, equipment design and organisational structures. It was only in understanding *how* to fight that Britain could hope to retain its hard-won warfighting edge and maintain the army in preparation for any future necessity. And it was only in understanding *how* to fight – and being structured, equipped and prepared to do so – that the army could hope to act as a strategic deterrent against any of the nation's potential enemies in Europe or elsewhere. Instead, because it had allowed its 1918 expertise to drain into the sand, by 1939 it was forced to go to war in Europe completely out of step with the Wehrmacht and the enemy's delivery of operational manoeuvre. The British Army was unable to fight a sustained, high-intensive war against a peer opponent in the same way that it had done – and triumphed – in 1918.

Contrary to common perception we have not found that the army was led during the inter-war years by donkeys, or that it failed to debate doctrine adequately. The period certainly experienced lots of talking, plenty of debate, much heat and (some) light. The problem was much more profound than this. Without understanding the importance of retaining what it had learned in 1918 the army simply forgot how to fight, or to be a little more generous was distracted by the pressure of empire and of severe financial stringency. In 1918 the sudden end to the fighting had the perverse effect of allowing the British quickly to forget what had delivered them victory. By 1919 the reasons for success in 1918 were quickly fading in the army's collective memory, as it dispersed to peace, deemed irrelevant in the fleeting exuberance of victory as khaki serge was exchanged for pinstripes and bowlers. The desire for 'hostilities only' officers to return to civilian life and for professional soldiers to put the trauma of an unusual and extraordinary war behind them was overwhelming.

The army had not forgotten how to garrison imperial outposts, nor how to undertake the essential and simple routines of soldiering, but it quickly forgot how to *warfight* on an intensive scale with a peer opponent. The features of warfare that had made it masters of the battlefield in 1918 were quickly neglected, and had to be relearned in the chaos, confusion and defeat of the early years of the Second World War. This was the army's fault, but with some mitigation from events outside its control. Whatever the multitudinous pressures on it during the inter-war years – public apathy, imperial overstretch and Treasury parsimony, among others – the army's first duty should have been to hang on to its inheritance, instead of neglecting it so egregiously in the years that followed the ending of the Great War.

What the British Army should have done after 1918 was to preserve the knowledge of how victory was achieved, and in the years following develop and codify it into something even more fearsome and formidable. It did not do so. Instead, by reverting to a regime based on 1914 it regarded anyone in its ranks who argued for the 1918 approach to be considered as an oddball, a fanatic, a heretic.

But no one retained responsibility for preserving the precious recipe that had led to such devastating success for the Allies in 1918. Why did this occur? Twelve Failures assert themselves:

1. There was no clear articulation of the *warfighting* role of the army. Instead, a policy of 'defending the empire' was adopted which was so general and unspecific as to be virtually useless as an anchor against which the British Army could be structured, trained and equipped for the next intensive or peer-on-peer war.

2. Consequently, the policy of defending the empire removed an incentive to consider the requirements for modern warfighting on the Continent as opposed to becoming in essence a gendarmerie.

3. This itself inculcated a lack of imagination in the War Office about the possibility of future war in Europe and stifled the conversations and innovation that were necessary to create an army able to refight 1918.

4. Indeed, there was no collective memory and record of 1918, with the first assessment of the Great War not being undertaken until 1932.

5. In any case, the army found itself stretched across a multitude of commitments across the globe, from Iraq, Ireland, India and Russia and beyond.

6. Zero-sum inter-service budgets meant that where one service gained (e.g. Royal Navy or Royal Air Force), the other, usually the army, lost.

7. No determined effort was undertaken by the government to decide what actual threats existed to British interests against which the British Army should be structured, what doctrine it therefore needed to build, what equipment it required to fight, and how best it should be trained.

8. The importance of a central philosophy of combined, joint operations did not exist. Equally, the army had no place within its structures to own, develop and promulgate doctrine.

9. Outside of the harmony forced on the army by war, peace reasserted its pre-war regimental tribalism and inter-service factionalism.

10. Individual officers and groups of lobbyists sought to prosecute their own ideas and agendas. Some were good and noble, others less so. Too often the single-issue lobbyists alienated their peers with fantasies about both the tank and the bomber.

11. The country, and army, were war weary: even committed professional soldiers found it a struggle to retain a professional interest in recent history that could have led to doctrinal development.

12. The ten-year rule, which removed the incentive to prepare for a serious war, resulted in the army being starved of funds.

But some of these points, such as lack of money or global dispersal, are merely excuses. It is an uncomfortable conclusion to reach, but on the evidence examined it is clear that the leadership of the British Army after 1918 fundamentally failed itself and the country it was charged with protecting, because it did not hold firm to those tactical, operational and doctrinal truths which it had mastered in 1918. It forgot the lessons of victory, and spent the following two decades in an intellectual, structural and doctrinal torpor. There was lots going on in these two decades, as we have demonstrated, but little was constructed on the basis of the clear successes and lessons of 1918. In fact, much of the doctrinal argumentation of the 1920s was a distraction. It didn't point back to 1918 and was deeply factional and tribalistic at heart. It wasn't a centrally co-ordinated debate owned and directed by an army leadership determined to construct a set of doctrinal truths around which the entire army (and armed forces) would happily coalesce, based on what had been learned and understood in 1918. That had all been ditched as irrelevant to future war. Instead, the army leadership appeared, for much of this time, to be detached observers as the government struggled with the conflicting issues of the economy, the demands of empire, Irish independence and an over-riding popular desire to avoid another war in Europe. This failure by the army's leadership as a whole to debate, decide and direct a future operational doctrine, led directly to its abysmal performance in Norway and France in 1940 and its failure to understand in North Africa in 1941 and 1942 the ingredients of battlefield success. The army was forced to learn the essentials of successful warfighting the hard way, again.

This book is about not just the failure of an army to prepare for war, but also the failure of both political and military leadership and dis-functionality between the two. It is the story of the decline of an army from the triumph of victory in 1918 to defeat in 1940. It is a cautionary tale for our own times. If there is ever a temptation by the British government to relax its attention to the need to ensure the army remains capable of fighting, when and where required, at the highest levels, then a glance back to the neglect of the 1920s and 1930s, and the humiliation of 1940, should be a sharp wake-up call. An army in being is a deterrent in itself, and not something to be cobbled together in a crisis when deterrence or diplomacy have failed.

What should the British Army have done after 1918 to prevent the disaster that was to befall the country two decades later? In addition to ensuring the amelioration of the Twelve Failures identified above, Britain should have:

1. Retained a core of its warfighting doctrine and capability from 1918, rather than dispersing it in a 'peace dividend'.
2. Articulated this knowledge as the basis of the British Army's new approach to war.
3. Constantly tested new approaches, ideas and concepts against this doctrine as the years went by, and found a home within the War Office's structure where innovation and doctrine could coalesce.

4. Undertaken a careful assessment of its potential enemies, against which the British Army might need to be deployed. It should then have constructed its warfighting structures around the enemies it expected it might have to fight, because if nothing else the army is an insurance policy which, to be effective, needs to be able to cover all risks.

After the early humiliation and the potential for defeat in the opening years of the Second World War, victory was finally achieved in 1945. However, the geo-strategic context of 1945 was markedly different from that of 1918. Although the Great War had been seen as the war to end war, the advancing tide of the Soviet Red Army from Stalingrad to Berlin redrew the map of Europe, ushering in the Cold War, while the atomic bombs dropped over Hiroshima and Nagasaki heralded the nuclear age. While the clamour for demobilisation was as loud in 1945 as in 1918, the demands of world policing saw the reintroduction of National Service in 1949 and a whole generation of civilians experiencing a brief time in uniform in the service of the Crown. Although nearly bankrupt in 1945, the Labour government still managed to introduce the Welfare State sketched out by the Beveridge Report of 1942 and finally repaid its war debt to the United States in 2006.

Although worldwide events were dynamic from 1945 to 1989 with further conflicts for the United Kingdom in Malaya, Dhofar, Cyprus, Kenya, Borneo, the Falklands, and the long-running Troubles in Northern Ireland, it was the Cold War in Europe that principally drove the defence agenda and kept the budget at around 5 per cent of Gross Domestic Product. As the major bridge between the United States and Europe, the Royal Navy was heavily committed above and below the surface of the Atlantic Ocean to keep open the sea lines of communication to NATO's dominant partner, while the British Army retained some 55,000 troops in four armoured divisions as part of NATO's Northern Army Group and the Royal Air Force was also largely forward-based in West Germany as part of the Second Allied Tactical Air Force. These conventional deployments were all conducted under the nerve-wracking nuclear umbrella of Mutual Assured Reduction. By the 1980s, with the West under the leadership of US President Ronald Reagan and British Prime Minister Margaret Thatcher and with increased spending on both conventional armaments and the highly experimental Inter-Continental Ballistic Missile Defence system, the strain of strategic military competition began to show on the political and economic stability of the Soviet Union. Despite the *perestroika* political movement for reform within the Communist Party of the Soviet Union and the associated openness of *glasnost* under General Secretary Mikhail Gorbachev, the cracks in the Berlin Wall that opened on 9 November 1989 led inexorably to the collapse of the Soviet Union two years later and the old flag of Russia being raised over the Kremlin on 26 December 1991. The Cold War was over, and an apparent New World Order had begun. The historian Francis Fukuyama declared – somewhat ambitiously – the end of history.

It was at this point that international leaders and their finance ministers in the West began to overlook the cautionary tale that the history of the 20th century might have taught them. With the Soviet Union gone and rump Russia apparently enfeebled, Western states eagerly embarked on military reduction and a peace dividend. In the United Kingdom, the 'Options for Change' exercise saw a major slashing of defence capability, beneficially coincidental to help ameliorate a significant economic downturn. The British Army was reduced from 155,000 to 116,000 soldiers, notwithstanding the first Gulf War of 1990–91 which many wishful thinkers regarded as something of an aberration. However, despite that war and the subsequent deployment of large parts of the armed forces to Bosnia from 1992 and then to Kosovo in 1999, the new Labour government of Prime Minister Tony Blair continued with the implementation of its Strategic Defence Review of 1997–98. As a piece of policy work, this was considered an honest review of the United Kingdom's defence policy and a progressive blueprint for future defence planning and expenditure. Endorsed by Tony Blair and the Chiefs of Staff, this review might have stood the nation in good stead for the future had the Chancellor of the Exchequer, Gordon Brown, fully funded its outcome. For his own reasons, he chose not to do so. The underfunding of the United Kingdom's defence capability began to show its deficiencies a year after with the second Gulf War of 2003, and the situation was then exacerbated by a protracted campaign in Iraq for the British Army lasting until 2009 and an even more intense one in Afghanistan lasting until 2014.

But domestic politics and financial austerity had already begun to undermine the reality of developments within Europe. The new British prime minister David Cameron in 2010 authorised a further reduction in the size of the British Army from 102,000 to 95,000 then another to 84,000 as a response to the economic downturn of 2008 and, notwithstanding the Russian annexation of Crimea in 2014, a further reduction was ordered by successive Conservative governments to 74,000. It is, perhaps, little wonder that Vladimir Putin, in power in the Kremlin in one appointment or another since 1999, sensed that the West, once united during the years of the Cold War, was ripe for challenge as he saw Western military capability being atrophied in Europe and diverted elsewhere in forays against militant Islamist movements in Iraq, Afghanistan and elsewhere in the Middle East and Africa. He may have taken some comfort when he witnessed the West's eventual disappointments in Iraq and Afghanistan, notwithstanding the Soviet Union's own failure in Afghanistan, but he will undoubtedly have been emboldened by the precipitate flight of the United States and NATO's undignified exit from Kabul in August 2021.

This is where this brief summary of recent events complements the longer analysis of the failings between the First and Second World Wars which is the principal subject of this book. If wrong decisions are made in the aftermath of a major conflict, like the First and Second World Wars and the stand-off that was the Cold War, then the United Kingdom, NATO and the West more generally stand in great danger, when faced by an expansionist challenger. In the 1930s, the liberal democracies of the West set their faces against rearmament and chose negotiation and appeasement as their response to Hitler's Germany. The war in Europe between Russia and Ukraine represents another moment when Western leaders, and those of the United Kingdom in particular, are faced with a dilemma to know how to act appropriately. They can choose to bury their heads in a Munich-like moment or otherwise embrace difficult and expensive choices, perhaps not popular in the short term but likely to be a sound investment in future security.

In 2022, accepting that Ukraine was not a member of NATO and therefore not under the umbrella of Article 5 of the North Atlantic Treaty, Western states, nevertheless, stood solidly in support of Ukraine as it fought to protect its independence, sovereignty and preferred way of life. But should it have come to this? Had the cautionary tale of the past not been heeded? Had the absence of a response to the rise of a dictator in Europe in the 1930s been mirrored by the hesitant response to the rise of a dictator in this decade? This is the challenge to Western governments in the third decade of the 21st century and to the United Kingdom, in particular, if it wishes to substantiate itself as Global Britain and as a major player on the world stage in the years to come. As a permanent member of the United Nations Security Council it is difficult to conceive how the United Kingdom can duck these responsibilities, but rising to them comes with a price. The Integrated Review of Security, Defence, Development and Foreign Policy published in March 2021 posited, inter alia, a geographic tilt towards the Indo-Pacific region and the prioritisation of new technologies and new ways of warfare; however, a brutal land war in Europe has proved to be a rude wake-up call. While the analysis behind the Integrated Review, largely confirmed by the refreshing of the review in March 2023, may have pointed to the way that the British government wished to act in the competitive age that it described, Vladimir Putin's violent assault on Ukraine took war back to its bloody basics. His military, having failed in a woefully inept attempt to make a lightning strike on Kyiv, had to resort to relentless pounding by rocket and tube artillery as a precursor to hapless Russian infantry being pressed forward to possess shattered territory. Meanwhile, supported by Western arms and ammunition, Ukrainian soldiers have demonstrated once again that Napoleon's dictum that 'The moral is to the physical as three is to one' is timeless.

So, what should be the response of the British government in the face of current events? The argument remains very strong that defence spending should rise to closer to 3 per cent than 2 per cent of Gross Domestic Product, much of which needs to be spent on a major investment in the United Kingdom's land power capability. Planned cuts to the size of the army should be questioned, as should decisions to reduce helicopter and tactical air lift capabilities. The Challenger 3 tank modernisation programme should be accelerated, and the numbers significantly increased from the paltry plan of just re-fitting 148 main battle tanks. Following the decision to remove the Warrior infantry fighting vehicle from service, it still remains vital that a suitable replacement is procured to enable armour and infantry to manoeuvre on the battlefield at a similar speed. Moreover, as the war in Ukraine has shown, rocket and tube artillery numbers need to be dramatically increased as does air defence capability, needed to meet conventional aerial threats and now that from armed drones. Underpinning all this must be a modern and secure communications system to enable agile command and control, all supported by robust logistics and adequate holdings of ammunition and other combat supplies. Without these enhancements and more, the United Kingdom will be unable to stand on the borders of Europe as a serious NATO partner and play its part in the future deterrence of further aggression from a newly aggressive Russia. Wishful thinking does not buy peace, but hard power does. A well-found army in being is a strong deterrent in its own right. It was absent in 1939 and disaster followed in 1940.

The history of the 1930s showed the folly of not acting in a timely manner while the draining away of the United Kingdom's military capabilities since the end of the Cold War shows a remarkable tendency to allow history to repeat itself. If this book is indeed a cautionary tale, its message needs to be carefully considered and acted upon without further delay. Defence expenditure is the insurance premium that any responsible government must pay to protect the security of its territory, its national interests and its people. The premium may have just gone up, but the alternative is a disaster. Ukraine is Europe's wake-up call. We must listen and act. We might not be given a second chance – again.

Appendix

Chiefs of the Imperial General Staff 1915–46

General Sir William Robertson (1860–1933)	23 December 1915	19 February 1918
Field Marshal Sir Henry Wilson (1864–1922)	19 February 1918	19 February 1922
Field Marshal Rudolph Lambart, 10th Earl of Cavan (1865–1946)	19 February 1922	19 February 1926
Field Marshal Sir George Milne (1866–1948)	19 February 1926	19 February 1933
Field Marshal Sir Archibald Montgomery-Massingberd (1871–1947)	19 February 1933	15 May 1936
Field Marshal Sir Cyril Deverell (1874–1947)	15 May 1936	6 December 1937
General John Vereker, 6th Viscount Gort (1886–1946)	6 December 1937	3 September 1939
General Sir Edmund Ironside (1880–1959)	4 September 1939	26 May 1940
Field Marshal Sir John Dill (1881–1944)	26 May 1940	25 December 1941
Field Marshal Alan Brooke, 1st Viscount Alanbrooke (1883–1963)	25 December 1941	25 June 1946

Notes

PROLOGUE

1 Gris Davies-Scourfield, *In Presence of My Foes: A Memoir of Calais, Colditz and Wartime Escape Adventures* (Barnsley: Pen and Sword, 2004), p. 19.
2 Hugh Sebag-Montefiore, *Dunkirk: Fight to the Last Man* (London: Viking, 2006), p. 230.

CHAPTER 1

1 Rawlinson's War Journal, 5 October 1918, GBR/0014/RWLN 1/9, Churchill Archives Centre, Churchill College, Cambridge.

CHAPTER 2

1 Duff Cooper, *Old Men Forget* (London: Rupert Hart-Davis, 1953), p. 83.
2 Gary Sheffield and John Bourne (eds), *Douglas Haig: War Diaries and Letters 1914–1918* (London: Weidenfeld and Nicolson, 2006), diary entry for Monday 29 July 1918.
3 IWM Sound Archive recording (Old Pals Productions, 1984) 16467.

CHAPTER 3

1 Field Service Regulations, Part I, Operations, 1909 (London, War Office, 1909); Field Service Regulations, Part I, Operations, 1909, Reprinted with Amendments 1912 (London, General Staff, War Office, 1912).

CHAPTER 4

1 *Infantry Training, The War Office* (London: His Majesty's Stationery Office, 1920).
2 Charles Carrington, *Soldier From the Wars Returning* (London: Hutchinson, 1965), p. 233.
3 James Beach (ed.), *The Division in the Attack – 1918, SS. 135* (Shrivenham: Strategic and Combat Studies Institute Occasional Paper, No 53, 2008), p. 15.
4 Corelli Barnett, *Britain and her Army* (London: Allen Lane, 1970), p. 409.

CHAPTER 5

1 Carrington, *Soldier From the Wars Returning*, p. 221.
2 Winston Churchill, *The World Crisis, Vol. 4., The Aftermath* (London: Thornton Butterworth, 1929), p. 35.
3 Ibid., p. 36.
4 Francis Tuker, *The Pattern of War* (London: Cassell, 1948) p. 13.
5 *British Journal of Psychiatry*, 8 October 2018.
6 Robin Prior and Trevor Wilson, *Command on the Western Front: The Military Career of Sir Henry Rawlinson 1914–1918* (Oxford: Blackwell, 1992), pp. 390–91.
7 Carrington, *Soldier From the Wars Returning*, p. 263.

CHAPTER 6

1 Quoted in Keith Jeffery, *Field Marshal Sir Henry Wilson: A Political Soldier* (Oxford: Oxford University Press), p. 271.
2 Bernard Montgomery, *The Memoirs of Field Marshal Montgomery* (London: Collins, 1958), p. 39.
3 Quoted in Jeffrey, *Field Marshal Sir Henry Wilson*, p. 243.
4 The National Archives, RECO 1/876.
5 Adrian Boyle, *Trenchard, Man of Vision* (London: Collins, 1962), p. 89.

CHAPTER 7

1 The National Archives, CAB 23/15/616A dated 15 August 1919.
2 The National Archives, CAB 24/132/19.
3 Alan Moorehead, *Montgomery* (London: Hamish Hamilton, 1946), p. 62.

4 David Lloyd George, *The War Memoirs of David Lloyd George*, 6 volumes (London: Nicholson & Watson, 1933–36), Volume IV, pp. 2110–11.

5 Cooper., *Old Men Forget*, p. 80.

6 J.F.C. Fuller, *The Army in My Time* (London: Rich and Cowan Ltd, 1935), p. 175.

7 Ibid., p. 174.

8 The National Archives, CAB 24/132/19 dated 10 January 1922.

9 Hastings Ismay, *The Memoirs of General the Lord Ismay* (London: Heinemann, 1960), p. 31.

10 The National Archives, CAB 24/132/19.

11 John 'Jack' Masters, *Bugles and a Tiger* (London: Cassell, 1956), p. 190.

CHAPTER 8

1 *Britain's Modern Army* (London: Odhams Press, 1941), p. 82.

2 Ibid.

3 Ibid., p. 97.

4 Quoted in Brian Bond, *British Military Policy between the Two World Wars* (Oxford: Clarendon Press, 1980), pp. 129–30.

5 The National Archives WO 33/1297 Report of the Committee on The Lessons of the Great War (October 1932).

6 Fuller, *The Army in My Time*, p. 179.

7 Quoted in Lavinia Greachen, *Chink* (London: Pan Macmillan, 1989), p. 92.

8 David French, '"An Extensive Use of Weedkiller": Patterns of Promotion in the Senior Ranks of the British Army, 1919–1939' in David French and Brian Holden Reid (eds), *The British General Staff: Reform and Innovation, 1890–1939* (London: Frank Cass, 2002), p. 166.

9 The National Archives, The Chief of Staff's Annual Report for 1926.

10 Fuller, *The Army in My Time*, p. 182.

11 Ibid., p. 179.

12 Ibid., p. 188.

13 Bond, *British Military Policy*, p. 71.

14 Ibid., p. 68.

15 Ibid., pp. 67–68.

16 Kennedy, *The Business of War* (London: Hutchinson, 1957), p. 18.

17 Ibid., p. 355.

18 Michael Howard, 'Leadership in the British Army in the Second World War', in Gary Sheffield (ed.), *Leadership and Command* (London: Brasseys, 1997), p. 124.
19 Fuller, *The Army in My Time*, p. 76.
20 Kennedy, *The Business of War*, p. 123.
21 Ibid., pp. 199–200.
22 Daniel Todman and Alex Danchev, *War Diaries 1939–1945: Field Marshal Lord Alanbrooke* (London: Weidenfeld & Nicholson, 2001), pp. 188, 193, 243.

CHAPTER 9

1 Brian Holden Reid, *War Studies and the Staff College, 1890–1930* (Camberley: Strategic and Combat Studies Institute, 1992), p. 13.
2 Clifford Kinvig, *Scapegoat: General Percival of Singapore* (London: Brasseys, 1996), pp. 91–92.
3 J.F.C. Fuller, *The Foundations of the Science of War* (London: Hutchinson & Co. Ltd, 1926).
4 Liddell Hart, *Memoirs* (London: Cassell and Company, 1965), Volume 1, p. 228.
5 Quoted in John Gooch, 'A Particularly Anglo-Saxon Institution: The British General Staff in the Era of the Two World Wars' in French and Holden Reid (eds), *The British General Staff: Reform and Innovation, 1890–1939* (London: Frank Cass, 2002), p. 198.
6 Ibid.
7 Tuker, *The Pattern of War*, pp. 46–47; 50–55.
8 Army Doctrine Publications (ADP), Land Operations, Parts 1–6 (HMSO, May 2022).
9 Field Service Regulations, Volume 1 (London: HMSO, 1923), p. 7.
10 Ibid., p. 4.
11 Bond, *British Military Policy*, p. 262.
12 An argument made cogently by the historian Lewis B. Namier in *Europe in Decay: A Study in Disintegration, 1936–1940* (London: Macmillan, 1940), pp. 18–25.

CHAPTER 10

1 J.F.C. Fuller, *The Reformation of War* (London: Hutchinson, 1923).
2 Corinthians 14:8.

3 Harold Winton, *To Change an Army* (Lawrence, Kansas: University Press of Kansas, 1988).

4 L.B. Oatts, *I SERVE: Regimental History of the 3rd Carabiniers (Prince of Wales' Dragoon Guards)* (Norwich: 1966), p. 246.

5 General Bill Jackson and Field Marshal Dwin Bramall, *The Chiefs: The Story of the United Kingdom Chiefs of Staff* (London: Brassey's, 1992), p. 133.

6 Fuller, *The Army in My Time*, p. 177.

7 Ibid., p. 188.

8 Winton, *To Change an Army*, p. 197

9 *Mechanized and Armoured Forces* (London: The War Office, 1929), pp 9–10 and *Modern Formations* (London: The War Office, 1931), pp. 9–15 (both the 'Purple Primers' written by Lieutenant Colonel Charles Broad.

10 Ibid.

11 Liddell Hart, *Memoirs*, Volume 1, p. 161.

12 The correspondence is contained in the Burnett-Stuart Papers, Liddell Hart Centre for Military Archives, King's College, London.

13 Liddell Hart Military Archives, Liddell Hart Papers, A2/2.

14 The fifth annual Haldane Lecture, as reported in *The Times*, 29 May 1933.

15 Liddell Hart, *The Decisive Wars of History* (London: Faber & Faber, 1929), pp.145–46. This was republished in 1967 as 'Strategy'.

16 Barnett, *Britain and her Army*, pp. 413–14.

CHAPTER 11

1 Liddell Hart Military Archives, Hastings Ismay Papers, III/4/12.

2 Moorehead, *Montgomery*, p.60.

3 The National Archives, CAB 24/198/148.

4 Charles Loch Mowat, *Britain Between the Wars 1918–1940* (London: Methuen, 1955), p. 422.

5 Winton, *To Change an Army*, p. 23.

6 Alan Milne, *Peace with Honour: An Enquiry into the War Convention* (London; Methuen, 1934). During the Second World War Milne nevertheless served in the Home Guard.

7 The National Archives, CAB 53/17 Imperial Defence Policy.

8 The National Archives, CAB 53/21 Imperial Defence Policy.

9   Neville Chamberlain, Prime Minister, 8pm BBC broadcast, 27 September 1938.
10 The National Archives, CAB 24/229/12.
11 The National Archives, CAB 23/234, CIGS Memorandum 28 October 1932.
12 *The Times*, 29 May 1933.
13 The National Archives, CAB 24/244/14.
14 The National Archives, CAB 24/247/64. Report of the Defence Requirements Sub-Committee 28 February 1934.
15 Walter Reid, *Neville Chamberlain, The Passionate Radical* (Edinburgh: Birlinn, 2021).
16 Brian Bond (ed.), *Chief of Staff: The Diaries of Lieutenant General Sir Henry Pownall* (London: Archon, 1973), Volume 1, p. 42.
17 Adrian Phillips, *Rearming the RAF for the Second World War: Poor Strategy and Miscalculation* (Barnsley: Pen and Sword, 2022).
18 The National Archives, PRO 30/69/620E, 1 March 1935.
19 Mowat, *Britain Between the Wars 1918–1940*, p. 542.

CHAPTER 12

1   This and other quotations from Matthew Halton are from *Ten Years to Alamein* (London: Drummond, 1944).
2   Leland Stowe, *Nazi Germany Means War* (London: Faber & Faber, 1933).
3   J.R. Kennedy, *This, Our Army* (London: Hutchinson & Co, 1935), p. 256.
4   All quotations from Shirer are from William J. Shirer, *Berlin Diary: The Journal of a Foreign Correspondent, 1931–1941* (New York: Alfred A. Knopf, 1941).
5   Robert Lyman, *Under A Darkening Sky* (New York: Pegasus Books Ltd, 2018), p. 291.
6   Wallace R. Deuel, *Hitler and Nazi Germany Uncensored* (*Chicago Daily News*, 1941).

CHAPTER 13

1   Cooper, *Old Men Forget*, p. 205.
2   The National Archives, CAB T161/1314/4 (Statement relating to defence, Command 5107).

3 Phillips, *Rearming the RAF for the Second World War.*
4 Barnett, *Britain and her Army,* p. 421.
5 Oatts, *I SERVE,* p. 247.
6 Ismay, *The Memoirs of General the Lord Ismay,* pp. 98–99.
7 Ibid., p. 101.

CHAPTER 14

1 D.R. Thorpe, *Eden: The Life and Times of Anthony Eden, First Earl of Avon 1897–1977* (London: Chatto & Windus, 2003), p. 175.
2 Iain Macleod, *Neville Chamberlain* (London: Frederick Muller, 1961), p. 224.
3 All quotations from Cowles are from Virginia Cowles, *Looking for Trouble* (London: Hamish Hamilton, 1941).
4 Ronald Blythe, *The Age of Illusion: England in the Twenties and Thirties, 1919–1940* (London: Faber and Faber Ltd, 1963), p. 229.
5 Halton, *Ten Years to Alamein,* p. 48.
6 Dorothy Thompson, *The Saturday Evening Post,* Indianapolis, 27 July 1935.
7 Speech by the Führer to the Commanders in Chief on 22 August 1939, United States Department of State, *Documents on German Foreign Policy: From the Archives of the German Foreign Ministry* (Washington, DC: US Government Printing Office, 1957–1964) Series D (1937–1945), The Last Days of Peace, Volume 7: August 9–September 3, 1939. Document 192, pp. 200–04.
8 Mowat, *Britain Between the Wars,* 1955.
9 Quoted in Ismay, *The Memoirs of General the Lord Ismay,* p. 124.

CHAPTER 15

1 Halton, *Ten Years to Alamein,* p. 56.
2 Kennedy, *The Business of War,* p. 13.
3 Halton, *Ten Years to Alamein,* p. 67.
4 See Timothy Harrison Place, *Military Training in the British Army, 1940–1944* (Abingdon: Routledge, 2000).
5 Brian Bond (ed.), *Chief of Staff, The Diaries of Lieutenant General Sir Henry Pownall,* Volume 1, 1933–1940, p. 294.
6 Kennedy, *The Business of War,* p. 2
7 Ibid.

8 Quoted in David Smurthwaite (ed.), *The Forgotten War* (London: National Army Museum, 1992), p. 17.
9 Kennedy, *The Business of War*, p. 29.
10 Ismay, *The Memoirs of General the Lord Ismay*, p. 109.
11 Quoted in Robin Prior, *Conquer We Must: A Military History of Britain 1914–1945* (London: Yale University Press, 2022), pp. 293–94.
12 The whole sorry story is brilliantly told in John Kiszely, *Anatomy of a Campaign: The British Fiasco in Norway, 1940* (Cambridge: Cambridge University Press, 2019).

CHAPTER 16

1 L.F. Ellis, *The War in France and Flanders* (United Kingdom History of the Second World War) (London: HMSO, 1953), pp. 11–12.
2 Bond (ed.), *Chief of Staff*, p. 315.
3 Ibid., p. 317.
4 Kennedy, *The Business of War*, p. 7.
5 Ismay, *The Memoirs of General the Lord Ismay*, p. 104.
6 Bond (ed.), *Chief of Staff*, p. 310.
7 Ibid., p. 311.
8 Ewan Butler, *Mason-Mac: The Life of Lieutenant General Sir Noel Mason-MacFarlane* (London: Macmillan, 1972), p. 116.
9 Todman and Danchev, *War Diaries*, p. 67.
10 See Edmund Ironside, *Ironside: The Authorised Biography of Field Marshal Lord Ironside* (Stroud: The History Press, 2018), pp. 350–57.
11 Montgomery, *The Memoirs of Field Marshal Montgomery*, p. 66.

CHAPTER 17

1 Davies-Scourfield, *In Presence of My Foes*, p. 49.
2 David Fraser, *Wars and Shadows: Memoirs of General Sir David Fraser* (London: Allen Lane, 2002), pp. 143–44. His account of the British Army in the Second World War is *And We Shall Shock Them: The British Army in the Second World War* (London: Hodder & Stoughton, 1983).
3 Final Report of the Bartholomew Committee on lessons to be learned from the operations in Flanders, The National Archives, CAB 106/220.

EPILOGUE

1 Brian Holden Reid, *A Doctrinal Perspective, 1988–1998* (Camberley: Strategic & Combat Studies Institute, 1998), p. 24.

2 The Army Field Manual, Volume 1, *The Fundamentals*, Part 1, 'The Application of Force' (HMSO: 1985).

3 John Kiszely, *The British Army and Approaches to Warfare Since 1945* (Camberley: Strategic & Combat Studies Institute, 1997), p. 8.

Suggestions for further reading

The literature on Britain during the inter-war period and of the political circumstances that led up to the Second World War is vast. That on the British Army is less so. With this in mind we felt that it would be wise to provide the reader with a selection of books – additional to those contained in the Notes – that would assist with further reading into the subjects we cover in this book, rather than slavishly listing all the many works we used in our research. This list is, therefore, not comprehensive, but comprises books we think readers will enjoy reading to explore the subject further.

Barnett, Corelli, *Britain and her Army* (London: Allen Lane, 1970)

Barr, James, *A Line in the Sand* (London: Simon and Schuster, 2011)

Becket, Ian and Simpson, Keith (eds), *A Nation in Arms: The British Army in the First World War* (Manchester: Manchester University Press, 1986)

Bidwell, Shelford and Graham, Dominick (eds), *Fire-Power: British Army Weapons and Theories of War 1904–1945* (London: George Allen and Unwin, 1982)

Blythe, Ronald, *The Age of Illusion: England in the Twenties and Thirties, 1919–1940* (London: Faber and Faber Ltd, 1963)

Boff, Jonathan, *Winning and Losing on the Western Front* (Cambridge: Cambridge University Press, 2012)

Bond, Brian, *British Military Policy Between the Wars* (Oxford: Oxford University Press, 1980)

Bond, Brian, *France and Belgium 1939–1940* (London: Davis Poynter, 1975)

Bouverie, Tim, *Appeasing Hitler* (London: Penguin, 2019)

Carrington, Charles, *Soldier from the Wars Returning* (London: Hutchinson, 1965)

Carver, Michael, *Britain's Army in the Twentieth Century* (London: Macmillan, 1998)

Carver, Michael, *Dilemmas of the Desert War* (London: Batsford, 1986)

Carver, Michael, *The Seven Ages of the British Army* (London: Harper Collins, 1986)

Charmley, John, *Duff Cooper* (London: Weidenfeld & Nicolson, 1986)

Clark, Lloyd, *Blitzkrieg* (London: Atlantic Books, 2016)

Cooper, Duff, *Old Men Forget* (London: Rupert Hart-Davis, 1953)

Corrigan, Gordon, *Mud, Blood and Poppycock* (London: Cassell, 2003)

Cottrell, Peter, *The Anglo-Irish War 1913–1922* (Oxford: Osprey, 2006)

Dennis, Peter, *Decision by Default: Peacetime Conscription and British Defence, 1919–39* (London: Routledge & Kegan Paul, 1972)

Dorney, John, *Peace After the Final Battle* (Dublin: New Island Books, 2013)

Edwards, Robert, *White Death – Russia's War on Finland 1939–40* (London, 2006)

Fraser, David, *And We Shall Shock Them* (London: Hodder & Stoughton, 1983)

French, David, *Raising Churchill's Army: The British Army and the War Against Germany 1919–1945* (Oxford: Oxford University Press, 2001)

French, David and Reid, Brian Holden (eds), *The British General Staff, Reform and Innovation 1890–1939* (London: Routledge, 2004)

Griffith, Paddy, *Forward into Battle* (Swindon: The Crowood Press, 1990)

Guderian, Heinz, *Achtung-Panzer!* (London: Weidenfeld and Nicolson, 1999)

Halton, Matthew, *Ten Years To Alamein* (London: Drummond, 1944)

Harris, Paul, *Men, Ideas and Tanks: British Military Thought and Armoured Forces, 1903–39* (Manchester: Manchester University Press, 2015)

Horne, Alastair, *To Lose a Battle, France 1940* (London: Macmillan, 1969)

Howard, Michael, *The Continental Commitment* (London: Maurice Temple Smith, 1972)

Jeffrey, Keith, *Field Marshal Sir Henry Wilson: A Political Soldier* (Oxford: Oxford University Press, 2006)

Kershaw, Robert, *Dünkirchen 1940: The German View of Dunkirk* (Oxford: Osprey, 2022)

Kinvig, Clifford, *Churchill's Crusade: The British Invasion of Russia 1918–1920* (London: Hambledon Continuum, 2006)

Kinvig, Clifford, *Scapegoat: General Percival of Singapore* (London: Brasseys, 1996)

Kiszely, John, *Anatomy of a Campaign: The British Fiasco in Norway, 1940* (Cambridge: Cambridge University Press, 2017)

Lukacs, John, *Five Days in London – May 1940* (Yale: Yale University Press, 1999)

Liddell Hart, Basil, *The Tanks: The History of the Royal Tank Regiment and its Predecessors, Heavy Branch Machine-Gun Corps, Tank Corps and Royal Tank Corps, 1914–1945*, 2 volumes (London: Cassell, 1959)

Lloyd, Nick, *The Western Front: A History of the Great War, 1914–1918* (London: Liveright Publishing Corporation, 2022)

Lloyd, Nick, *The Hundred Days: The Campaign that Ended the First World War* (London: Basic Books, 2014)

Mallinson, Allan, *Too Important for the Generals: Losing and Winning the First World War* (London: Bantam, 2016)

Mallinson, Allan, *The Making of the British Army* (London: Bantam Press, 2009)

Mearsheimer, John, *Liddell Hart and the Weight of History* (New York: Cornell University Press, 1988)

Mowat, Charles, *Britain Between the Wars 1918–1940* (London: Methuen and Co Ltd, 1955)

Overy, Richard, *The Morbid Years: Britain and the Crisis of Civilisation, 1919–1939* (London: Allen Lane, 2009)

Phillips, Adrian, *Rearming the RAF for the Second World War* (Barnsley: Pen and Sword, 2022)

Phillips, Adrian, *Fighting Churchill, Appeasing Hitler* (London: Biteback, 2019)

Prior, Robin, *Conquer We Must* (London: Yale, 2022)

Prior, Robin, *Command on the Western Front: Military Career of Sir Henry Rawlinson, 1914–18* (London: Blackwell, 1992)

Reid, Walter, *Douglas Haig, Architect of Victory* (Edinburgh: Birlinn, 2011)

Saunders, Tim, *Arras Counter-Attack 1940* (Barnsley: Pen and Sword, 2018)

Sheffield, Gary, *Douglas Haig: From the Somme to Victory* (London: Aurum, 2016)

Sheffield, Gary, *Forgotten Victory: The First World War: Myths and Realities* (London: Headline, 2001)

Sheffield, Gary and Gray, Peter (eds), *Changing War: The British Army, the Hundred Days Campaign and the Birth of the Royal Air Force, 1918* (London: Bloomsbury, 2013)

Shirer, William, *The Rise and Fall of the Third Reich* (New York: Secker & Warburg Ltd, 1959)

Shirer, William, *Berlin Diary: The Journal of a Foreign Correspondent, 1934–1941* (London: Hamish Hamilton, 1942)

Stowe, Leland, *Nazi Germany Means War* (London: Faber & Faber, 1933)

Strohn, Matthias, *1918 Winning the War, Losing the War* (Oxford: Osprey, 2018)

Terraine, John, *White Heat: The New Way in Warfare* (London: Sidgwick & Jackson Ltd, 1982)

Terraine, John, *To Win A War: 1918, the Year of Victory* (London: Sidgwick & Jackson, 1978)

Thompson, Julian, *Dunkirk: Retreat to Victory* (London: Arcade Publishing, 2015)

Thurlow, Dave, *Building the Gort Line* (Solihull: Helion and Company Ltd, 2019)

Townsend, Charles, *The Partition: Ireland Divided, 1885–1925* (London: Allen Lane, 2021)

Townsend, Charles, *The Republic: The Fight for Irish Independence, 1918–1923* (London: Allen Lane, 2013)

Travers, Tim, *The Killing Ground: The British Army, the Western Front and the Emergence of Modern War 1900–1918* (London: Harper Collins, 1987)

Trythal, Anthony, *Boney Fuller: The Intellectual General* (London: Cassell, 1977)

Tuker, Francis, *Approach to Battle: A Commentary; Eighth Army, November 1941 to May 1943* (London: Cassell, 1963)

Tuker, Francis, *The Pattern of War* (London: Cassell, 1948)

Winton, Harold, *To Change an Army: General Sir John Burnett-Stuart and British Armored Doctrine, 1927–1938* (Lawrence, Kansas: University Press of Kansas, 1998)

Ziegler, Philip, *Between the Wars* (London: Maclehose Press Quercus, 2017)

Index